A Laboratory Manual for

INTRODUCTION TO ENVIRONMENTAL SCIENCE

First Edition

Dawn M. Ford, PhD
UTC Academic Affairs

Bradley Reynolds, EdD
UTC Biology, Geology, and Environmental Science

With contributions from:
Jennifer Boyd, PhD –UTC Biology, Geology, and Environmental Science
David Stanislawski, PhD - Chattanooga State Chemistry Department
Thomas Wilson, PhD – UTC Biology, Geology, and Environmental Science

Kendall Hunt
publishing company

Cover image © Shutterstock, Inc.

www.kendallhunt.com
Send all inquiries to:
4050 Westmark Drive
Dubuque, IA 52004-1840

Copyright © 2014 by Kendall Hunt Publishing Company

Revised printing 2016

ISBN 978-1-4652-9512-5

Printed in the United States of America

TABLE OF CONTENTS

The Scientific Method

> *"Science is nothing but trained and organized common sense, differing from the latter only as a veteran may differ from a raw recruit: and its methods differ from those of common sense only as far as the guardsman's cut and thrust differ from the manner in which a savage wields his club."*
>
> Thomas H. Huxley - *English biologist (1825 - 1895)*

Student Learning Outcomes

1. To learn about the scientific method and start thinking about how to apply it.
2. To formulate a hypothesis and generate data supporting or disproving it.
3. To work with a living organism called a planarian.
4. To learn the basics of how a formal lab report is constructed.

Discussion:

The Scientific Method

Science is best defined as organized knowledge. It's a way of thinking. At the core of every science is the scientific method. If you were to ask a scientist to define the scientific method, the scientist's definition would depend to some extent on his or her discipline. An environmental scientist's definition might differ slightly from a chemist's or a sociologist's definition. Regardless of the discipline, the heart of the scientific method is experimentation and the collection and treatment of data.

The scientific method starts with ***observation***. Observation is using the senses - not just sight, but also taste, touch, smell, and sound - to take note of some occurrence or phenomenon (OF COURSE, YOU SHOULD NEVER TASTE ANYTHING IN LAB). The scientist then starts to formulate research questions about the observed phenomenon in his or her mind and comes up with a possible explanation for why the phenomenon is the way that it is (Reece, Taylor, Simon, & Dickey, 2015). For example, you may observe that the sky is blue. You may then formulate a possible explanation by stating that the blue color is caused by air molecules scattering blue light from the sun. This possible explanation of the observed phenomenon is called a ***hypothesis***. The hypothesis is based upon the observations (Reece, Taylor, Simon, & Dickey, 2015).

After developing a hypothesis, the scientist will attempt to test its validity through a series of tests. This is called ***experimentation***. If we continue with our example, experiments would be designed to determine the true nature of the way air molecules scatter the light of the sun. You might test the way blue light is scattered versus the way red light is scattered. It is during the process of experimentation that data are collected. Data can be quantitative (numeric) or qualitative (word-based and descriptive). Following the collection of data, the scientist must examine the data and make ***generalizations***. When the scientist makes generalizations, he or she is looking for relationships and patterns in the data to either support or disprove the hypothesis (Botkin & Keller, 2010). Often the data are displayed in the form of a graph. A graph is simply data in picture form.

From the results of the experiment and the generated data, a scientist can then form his or her ***conclusions***: Was the hypothesis supported or disproved? If the hypothesis is disproved, it may be necessary to restart the process and generate a new hypothesis to explain the observed phenomenon. If the hypothesis is supported, additional experimentation should be done. If, after much experimentation, a hypothesis is supported over and over again, that hypothesis may eventually be classified as a ***theory***. A theory has attained the highest level of scientific certainty possible (Botkin & Keller, 2010).

To review, the steps of the scientific method are as follows: Make Observations/formulate Questions, formulate a hypothesis, experiment and collect data, evaluate the data and make generalizations; draw conclusions, and perform additional tests and/or reformulate the hypothesis.

The Formal Lab Report

In this book, you will find data sheets on which you are to write down your observations, record your data, and perform your calculations. Scientists, however, are often required to keep a traditional laboratory notebook or field journal. A lab notebook is a book with blank pages on which the scientist writes down the details of his or her experiment along with the results and conclusions. When writing in a lab notebook, certain guidelines must be followed and certain sections must be included. A properly written lab notebook will make it possible for another scientist to follow the previous experimenter's work and to duplicate that work exactly. For this reason, a lab notebook must always be thorough, detailed, and complete.

In this laboratory class and in others, you may be asked to write a formal laboratory report even if you are not asked to keep a traditional laboratory notebook. If you are asked to turn in a formal lab report in this class or in any other, it should also be thorough, detailed, and complete. Your formal lab report should be typewritten and it should include the following sections:

The **Title** should consist of a brief yet descriptive title for the report, located on a separate title page.

The **Introduction** should include the purpose of the experiment. What is your goal? What do you hope to determine (research question)? This section should also include relevant background information on the topic. Once the topic of study has been introduced, the hypothesis for the experiment should be stated toward the end of the Introduction section.

The **Materials** section should list all of the equipment and supplies used by the scientist to perform the experiment.

The **Procedures** section should include explicit detail on how the experiment was performed and the data collected. Provide enough detail that another scientist could use your report to repeat the experiment. A diagram or diagrams may be appropriate. Include the date, time, duration, and location of the experiment with the procedure.

In the **Results** section, you should include all qualitative observations and all quantitative data produced in the experiment. Make sure to include all of the data in logically constructed, properly-labeled tables and/or figures such as graphs. Include any necessary calculations in the results section after the data tables. If appropriate, calculate statistics in order to better describe the uncertainty of your measurements (the average range, the variance, the standard deviation, etc.)

In the **Discussion** section, the results should be actively analyzed and interpreted. It may be appropriate to compare your results to similar studies performed by other experimenters. After discussing your findings, be sure to state whether the hypothesis was supported or rejected. If it is supported, suggest additional means of testing the same hypothesis. If it is rejected, suggest how the hypothesis might be changed or modified. In either case, if appropriate, explain how the findings of your study are important and how your results might be useful in the future to other scientists and researchers. What are the practical applications of your work? An error analysis should be included within the discussion to identify and explain any relevant sources of error or bias that might have affected the outcome of your experiment. The discussion section is the most important part of the formal lab report. It is here that you show that you understand the data and that you explain the relevance of the data and the experiment.

The **Conclusion** should be a brief one to two sentence statement saying what is known conclusively as a result of the experiment. For example, a conclusion statement for next week's experiment might read as follows: *the pipette calibrated delivers an average aliquot volume of 4.97 mls with an average deviation of ± 0.02 mls.*

The References section should include a list of all sources (articles books, webpages, etc.) cited in the report in the correct format and in alphabetical or numerical order. In general, environmental scientists should cite their papers in the **Name-Year** format according to the *Council of Science Editors' (CSE) Scientific Style and Format.*

Today's data sheet is written and organized like an abbreviated formal lab report. You will be required to construct a hypothesis, make qualitative observations, collect and record quantitative data, discuss the data, and formulate a conclusion.

References:

Botkin, D.B. and Keller, E.A.. (2010). *Environmental Science – Earth as a living planet* (8th ed.). Hoboken, NJ: John Wiley and Sons.

Carolina Biological Supply Company. (2006). Carolina protozoa and invertebrates manual. Burlington, North Carolina. Available at http://www.carolina.com/teacher-resources/Document/protozoa-invert-care-handling-instructions/tr10466.tr.

Collins, L.T. and Harker, B.W. (1999). Planarian behavior: A student designed laboratory exercise. *In* Tested studies for laboratory teaching, 20, pp. 375-379. Proceedings of the 20th Workshop/Conference of the association for Biology Laboratory Education (ABLE). Available at http://www.ableweb.org/volumes/vol-20/mini8.collins.pdf.

Thomas Huxley Quotes, Quotes.net. (n.d.) Retrieved from http://www.quotes.net/quote/1555.

Reece, J.B., Taylor, M.R., Simon, E.J., Dickey, J.L., and Hogan, K.A. (2015). Campbell Biology: Concepts and Connections. Benjamin Cummings: San Francisco.

SCIENTIFIC METHOD DATA SHEET

Name: _____

Title: Investigating the Scientific Method

Purpose: To expose a planarian to light and to observe the reaction of a planarian to light, thereby gaining insight into the proper application of the scientific method.

Background and Introduction: Planarians are non-parasitic flatworms commonly found in freshwater environments (Figure 1). They move by crawling over secreted mucous and by the action of tiny hair like structures called cilia. They feed in part on organic debris and therefore play a role in natural decomposition. The mouth is located on the lower middle-part of the organism. The mouth is used not only for the intake of food but also to discharge waste. If a planarian is cut in two, it will often regenerate, producing two whole organisms (Carolina, 2006). Planarians are generally light sensitive. They can detect light through pigmented eyespots. If an organism moves toward a light source, it is referred to as positively phototactic. If an organism moves away from a light source, it is referred to as negatively phototactic (Collins & Harker, 1999).

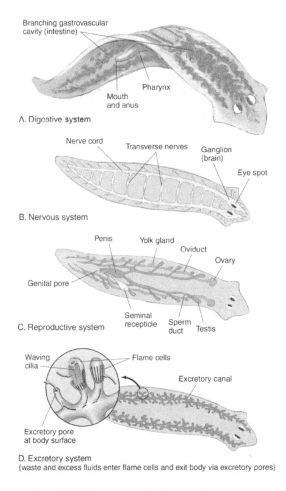

© Kendall Hunt Publishing Co.
Figure 1. Planarian anatomy.

Given the background information about planarians and the purpose of the experiment, develop a hypothesis for your study. The hypothesis for this experiment is as follows:

_____

_____

_____

Procedure: First, expose the planarian to light and note and record the response. Repeat several times. Continue the experiment by using black construction paper to divide a Petri dish into halves. The covered half will represent the 'dark side' and the uncovered half the 'light side.' Position a lamp or flashlight above the dish. Position the planarian in the exact middle of the dish and turn on the lamp. Use the lamp to directly illuminate the uncovered half. Observe the behavior of the planarian.

Record in the table the planarian's position every 30 seconds for 10 minutes: **dark** when the flatworm is on the dark side and **light** when the flatworm is on the light side.

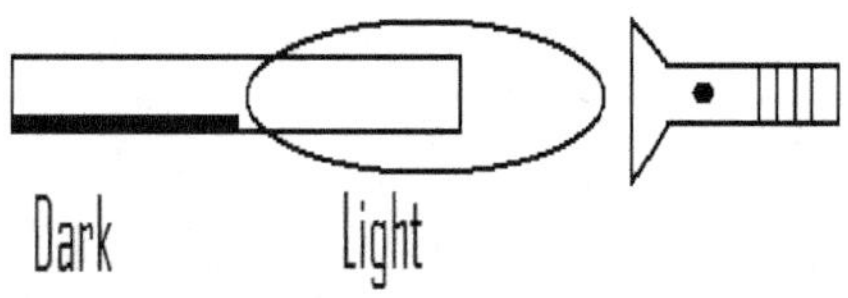

Dark   Light

Figure 2. Position of flashlight and petri dish.

Results: Quantitative

Time	Light or Dark	Time	Light or Dark
0.5 minutes		5.5 minutes	
1.0 minutes		6.0 minutes	
1.5 minutes		6.5 minutes	
2.0 minutes		7.0 minutes	
2.5 minutes		7.5 minutes	
3.0 minutes		8.0 minutes	
3.5 minutes		8.5 minutes	
4.0 minutes		9.0 minutes	
4.5 minutes		9.5 minutes	
5.0 minutes		10 minutes	

Calculations (Calculate percentage of time spent in the light and the percentage spent in the dark):

Results: Qualitative (Description of Planarian/Planarian's Behavior Before/After Light Exposure):

_____

_____

_____

Discussion: Analyze, interpret, and explain the data. Show how the data relate to the hypothesis:

_____

_____

_____

_____

6

_____
_____
_____

Conclusion: Write a brief 1 to 2 sentence statement saying what you know for sure as a result of the experiment. For this experiment state whether the planarian was mostly in the light or mostly in the dark and whether the planarian is positively phototactic or negatively phototactic.

_____
_____

An Introduction to Environmental Science and
Basic Laboratory Skills

"I am among those who think that science has great beauty. A scientist in his laboratory is not only a technician: he is also a child placed before natural phenomena which impress him like a fairy tale."

Marie Curie - *French (Polish-born) chemist & physicist (1867 - 1934)*

Student Learning Outcomes:
1. To become more familiar with the highly interdisciplinary field of environmental science.
2. To better understand the uncertainty associated with scientific measurement.
3. To read a graduated cylinder and how to use a volumetric pipette.
4. To read a balance.
5. To record measurements in the laboratory correctly, with the proper units and the correct number of significant figures.

Materials:
Five graduated cylinders filled with water and covered in parafilm, laboratory, thermometers, beakers, ice, balances with a labeled set of weighing unknowns, 5.00 ml pipettes with rubber pipette bulbs, calculators

Discussion:

What is Environmental Science?

Environmental science is most simply defined as the highly interdisciplinary study of the environment and the impact of human beings upon it. The ***environment*** includes both living and non-living components. Living components include the other organisms with which human beings interact. Non-living components include the air, water, and soil on which all life depends (human or otherwise). When we consider the environment, we must of course remember not only the natural world (lakes, rivers, forests, grasslands, etc.), but also the man-made world of cities, urbanized areas, and agricultural systems (Botkin & Keller, 2010). Over seven billion people now inhabit both worlds. Of course, such a massive global population puts a tremendous strain on the earth. At the same time, human technology is constantly increasing and expanding, allowing us to use the earth's resources for our own benefit. Obviously, when the sheer size of the human population is combined with the immense power of human technology, human beings have the potential to influence and change the environment in ways never before thought possible.

When the man-made world collides with the natural world, environmental problems are sometimes created. Environmental problems created by humans are incredibly complex and one often leads into another. Environmental problems typically revolve around human population growth, the earth's natural resources, and environmental pollution and degradation.

We use science to better understand environmental problems and to find solutions. Science is simply organized knowledge. It is a dynamic way of thinking that relies heavily on qualitative observations and on quantitative measurements (Botkin & Keller, 2010). It has already been stated that environmental science is interdisciplinary, meaning that it draws on a variety of disciplines to accomplish its goals: law, policy, philosophy, geography, ecology, biology, chemistry, geology, etc. In spite of its interdisciplinary nature however, environmental science is still first and foremost an experimental

9

laboratory and field science. **The purpose of the lab experiment that follows is to familiarize you, the student, with some basic scientific equipment like graduated cylinders and balances and to equip you with a foundation for making and recording data measurements in environmental science.**

How to Read a Graduated Cylinder and the Importance of Significant Figures

Beakers are used only for crude measurements. If a beaker reads 300 milliliters (mls) of a liquid, it *may* actually contain 300 mls, but then again it may contain 280 mls or 320 mls. Graduated cylinders, on the other hand, are for precise work and can be trusted to give a true volume reading. When reading a graduated cylinder, one should be eye level with the meniscus. The meniscus is the curve of the liquid in the graduated cylinder. In addition to being eye level with the meniscus (to avoid parallax which would produce a faulty reading), one should note the volume at the bottommost part of the meniscus

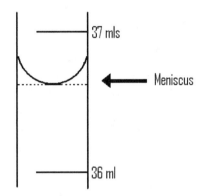

Figure 1. The meniscus in this graduated cylinder is between 36 and 37 mls.

In Figure 2, the bottom of the meniscus falls between 36 and 37 mls. Of this much, we are certain, but where exactly does it fall between 36 and 37 mls? If one counts the gradations, or lines of division, on the body of the cylinder, it is obvious that it falls between 36.6 and 36.7 mls. This is also sure and certain. Where exactly does it fall between 36.6 and 36.7 mls? We must at this point mentally divide the distance between 36.6 and 36.7 up into ten equally sized units and *estimate* where the bottom of the meniscus lies. It appears to lie about 3 tenths of the way between 36.6 and 36.7. Therefore, we would report the volume of the liquid in the graduated cylinder as 36.63 mls. We are certain and sure about the first three numbers in the measurement, but there is some uncertainty associated with the last number. Another scientist may look at the meniscus and call it 36.*62* mls or 36.*64* mls. Either is acceptable, because again the last number is an ***approximation***, an estimate, and because different people tend to estimate distances in slightly different ways. Human error is built into the last digit. The preceding example illustrates an important rule when making a measurement in the laboratory. **In general, read as many numbers as you are certain of and estimate the last digit.** Now consider the following graduated cylinder. Notice that it does not have as many gradations.

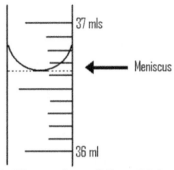

Figure 3. The meniscus falls at 36.6 mls.

Clearly, the bottom of the meniscus lies between 36 and 37 mls. We must estimate exactly where it lies between the 36 and 37. It appears to fall on 36.6 mls. This time we are sure about the first two numbers, but we have to estimate the third digit. **We have read as many numbers as we are certain of, and have estimated the last one.** Because this second graduated cylinder has fewer gradations, it is not as precise as the first one, and it has given us fewer *significant figures*. Significant figures reflect the uncertainty of our measuring device and represent how sure we are about a measurement. To report more significant figures than an instrument is capable of giving is to misrepresent how sure we are about the measurement (be it a graduated cylinder, a thermometer, or a balance). On the other hand, to leave off a digit and 'cheat' ourselves of a significant figure is to say that we are not as sure about a measurement as we could be. Whatever the instrument, **we should always report laboratory measurements with the proper number of significant figures and with the proper unit.** The proper number of significant figures is of no value without the proper unit and vice versa.

References:

Botkin, D.B. and Keller, E.A.. (2010). *Environmental Science – Earth as a living planet* (8th ed.). Hoboken, NJ: John Wiley and Sons.

Cochran, J., and Stanislawski, D. (2002). *General chemistry I laboratory and activity manual*. Wiley Custom Services: USA.

Quotations Page. (2013). Marie Curie. Retrieved from http://www.quotationspage.com/quote/34022.html

Raven, P., Hassenzh, D.M, and Berg, L. (2012). *Environment* (8th ed.). Hoboken, NJ: John Wiley and Sons.

The Microscope

Student Learning Outcomes:

1. To know the parts and to use the stereomicroscope.
2. To know the parts and use the compound microscope.
3. To make a wet slide.
4. To stain a slide.
5. To sketch a plant cell, an animal cell, a water flea, and other specimens.

Materials:

Stereomicroscope, a compound microscope, a water dropper, flat slides, cover slips, an onion, forceps, methylene blue or other biological stain, flat toothpicks, pond water, small Petri dishes, water fleas

Discussion:

You have already familiarized yourself with the graduated cylinder, the thermometer, the pipette, and the balance and learned how to take accurate measurements with several different pieces of basic laboratory equipment. Today, you will familiarize yourself with another important piece of laboratory equipment: the microscope. The microscope is an optical instrument that utilizes lenses to produce magnified images of extremely small objects. **Every environmental scientist should be proficient with the microscope.**

There are several different types of microscope. In today's lab, we will use both the stereomicroscope and the compound microscope. Regardless of the type of microscope used, chances are it is very expensive, so the first thing you should know is the proper way to carry it. Always carry the microscope with one hand under the base and the other hand grasping the arm of the microscope. Your instructor will demonstrate when he or she familiarizes you with the different parts of the microscope.

The *stereomicroscope* is used for viewing an object under low magnification. Typically, the stereomicroscope has a pair of 10X eyepieces called ocular lenses and a zoom lens. The ocular lenses along with the zoom lens make it possible to see a three-dimensional image of the object of interest. Stereoscopes are used to examine larger objects in great detail. Objects that are too big to be placed on the stage of a compound microscope are appropriate for stereomicroscopes. Stereomicroscopes are used for larger objects because the lower magnification of a stereomicroscope allows for a wider field of view.

TO USE A STEREOMICROSCOPE, simply place the object to be examined on the stereomicroscope's stage and illuminate the object with an external light source. It should be noted that on some stereomicroscopes the light source is built into the instrument. As you look through the eyepieces, bring the object into focus with the focus knob, usually found on the side of the scope. Use the zoom lens to either move in very close on the object or to draw back away from it.

The *compound microscope* is used for the high magnification of extremely small specimens like cells and microscopic life forms that could never otherwise be seen using only the unaided eye. The compound microscope consists of an eyepiece lens (the ocular lens) and an objective lens at opposite ends of the main part of the instrument's tubular body. Both lenses contribute to the total magnification of the object. The eyepiece lens magnifies 10X. The scanning power objective lens typically magnifies 4X, the low power objective lens typically magnifies 10X, and the high power objective lens normally magnifies 40X. The objective lenses will be labeled with the power of their magnification and color coded: 4X in black writing, 10X in green, and 40X in yellow. To calculate the total magnification, multiply the power of the eyepiece lens by the power of the objective lens being used. For example, if you are looking at a specimen with the high power lens, you would need to multiply 10 x 40 for a total magnification of 400X.

TO USE THE COMPOUND MICROSCOPE, turn on the microscope light. Move the revolving nosepiece so that the lowest power objective lens is in position. Now place the slide on the stage and secure the slide using the stage clips. Make sure the slide is placed on the stage so that the specimen to be viewed is directly under the center of the objective lens and in the beam of light. Adjust the diaphragm to change the intensity of the beam. Less light will allow for the observation of more detail. Look through the eyepiece and use the large knob (the *coarse adjustment*) to bring the object into focus as much as possible. Then, use the smaller knob (the *fine adjustment*) to bring the specimen into even sharper focus. To view the same specimen under higher magnification, just move the revolving nosepiece and select a higher power objective lens. Again use the fine adjustment knob to bring the specimen into sharp focus.

Note: Some microscopes have oil immersion lenses with a magnification of 100X. These lenses should not be used without oil for lubrication. Trying to use the oil immersion lens without oil will damage the lens.

References:

English Poetry II: From Collins to Fitzgerald. Vol. XLI. The Harvard Classics. New York: P.F. Collier & Son, 1909–14.

Kingsley, C. (1855). *Glaucus; or, the wonders of the shore.* Cambridge: Macmillan Company.

Virginia Community College. (n.d.). Protozoa – Lesson 6. Retrieved from http://water.me.vccs.edu/courses/env108/lesson6_2.htm.

MICROSCOPE DATA SHEET

Name: _____

A. Labeling - Listen as your instructor explains the various parts of the stereomicroscope and the compound microscope. Note the different functions of the various parts and label the pictures of the stereomicroscope and the compound microscope appropriately.

For the stereomicroscope, label the ocular lenses, the base, the stage, the arm, the zoom knob, and the focus knob.

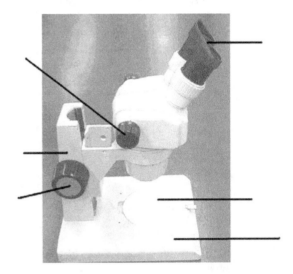

For the compound, label the ocular lenses, the objective lenses, the base, the stage, the arm, the light, the course adjustment knob, and the fine adjustment knob.

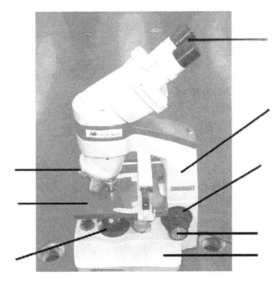

B. Using the Stereomicroscope - Examine a Rock, an Insect, a Flower, etc.

Detailed Sketch (with a pencil only):

Your instructor will provide you with a variety of objects. Remember that a stereomicroscope is used to examine slightly larger objects, so that the entire object is easily located within the wider field of view.

Sketch at least one of the objects in great detail in the space provided. You may opt to examine the **Japanese beetle**, a destructive invasive accidently introduced to the U.S. in a shipment of iris bulbs. Use your powers of observation. Your grade will be based in part on effort and on how thorough your sketches are, so take these drawing exercises seriously.

C. Using the Compound Microscope - A Water Flea

Observe a water flea by following these steps:
1. Acquire a water flea. Water fleas are important in environmental science because they are excellent ecological indicators and often used in environmental monitoring.
2. Place the organism on a slide with a dropper.
3. DO NOT use a cover slip.
4. Carefully draw away excess water with a paper towel.
5. Observe the water flea's movement under the **compound microscope**.
6. Sketch the organism. It is ossible to look through a water flea and see its beating heart and moving blood cells.

Detailed Sketch (with a pencil only): The box is on the next page.

D. Using the Compound Microscope - Looking at Cells and at a Wet Mount of Pond Water

As environmental scientists, we are not interested in onion cells per se; we are interested in becoming competent with the microscope and in learning how to prepare slides. So to practice, observe a plant cell by following these steps:

1. Cut a small piece of onion and peel off a piece of the onion skin with forceps. This is the onion's surface layer of cells.
2. Place the onion skin in the middle of a clean slide and flatten it with your finger.
3. Add one drop of water.
4. Cover the onion skin by lowering a cover slip gently over the drop at an angle, with one edge of the cover slip touching the flat slide first, allowing the water to spread out. Make sure no air bubbles form.
5. Stain the sample by adding a single drop of methylene blue on one side of the cover slip. On the opposite side, use a paper towel to draw the stain across the sample (be sure to soak up any excess stain). Staining makes the viewing of details somewhat easier.
6. Place the slide on the **compound microscope** stage and observe at low magnification. Draw what you see in the space provided. Label the nucleus, cell wall, and cytoplasm. The rounded *nucleus* contains the DNA. It is the control center of the cell. The *cell wall* is the outermost layer of a plant cell. The *cytoplasm* is the living material inside the cell, other than the nucleus. Repeat this process at a higher magnification.

onion (plant) cell	onion (plant) cell
Magnification _____	Magnification _____

Collect a sample of pond water or use the sample provided by the instructor. Remove a sample of the water from the bottom of the jar and prepare a wet-mount slide (with a single drop of pond water and a cover slip). Try to find tiny pond animals or animal-like protozoa or plant-like algae under the **compound microscope**. Sketch what you see. Use the identification key provided and try to identify some of the various organisms. If you have trouble, do not be discouraged. There are literally thousands of different species of protozoa alone (Virginia Community College, n.d.)!

Magnification _____	Magnification _____

Information Literacy

"I myself spent hours in the library as intimidated and embarrassed as a famished gourmet invited to a dream restaurant where every dish from all of the world's cuisines, past and present, was available on request." - Luigi Barzini

"But what can a man see of a library being one day in it?" - James Boswell

Student Learning Outcomes
1. To locate peer-reviewed journal articles in the field of environmental science.
2. To know sources of current and reputable science information.
3. To cite references using CSE style.

Discussion:
Using the Library and Citing Sources

When writing a formal lab report (or keeping a lab notebook), it is often necessary to find and make use of current and relevant sources when gathering background information. It is also often helpful to find similar studies performed by other scientists in order to compare their findings with your own (Cox, 2001). It is therefore extremely important that an environmental scientist be able to effectively find, use, and properly cite the resources at his or her disposal. When researching sources for a formal lab report or for a traditional lab notebook, the library is an excellent place to begin. The library on your campus will allow you access to a wide variety of resources.

A library's **online catalog** can normally be accessed anywhere by any computer (See www.lib.utc.edu) The online catalog allows students to search the library's collection of resources, including books, journals, magazines, the somewhat antiquated microforms and microfiche, and audiovisual materials. If a student wishes to use a source not owned by the campus library, interlibrary loan is always an option. If a student wants a particular book or journal article on global climate change for example, and his or her campus library does not own it, the campus library may be able to 'borrow' the book or article from another library that does own the title.

Print indexes, although antiquated, are also places to find information. Print indexes are available in most libraries. They are essentially books of references. Examples of common print indexes include Environmental Abstracts (GF1 .E553) and the Biological and Agricultural Index (S1 .A36). Print indexes generally contain information about older journal articles. Newer, more up-to-date journal articles can be found through **electronic indexes**. Examples of electronic indexes include such databases as Infotrac Onefile, Agricola, Basic Biosis, and Environmental Sciences & Pollution Management.

The Internet

The internet, or World Wide Web, is an excellent source of information; however, one should always be cautious when using the internet. Just because information is posted on the internet does not necessarily mean that the information is reliable or accurate. Anyone could potentially post misinformation on the internet (your instructor may intentionally post some misinformation about his or her own career and background just to prove this point). For this reason, only use information from reputable websites. Reputable websites are normally websites associated with entities such as federal and state government and universities and colleges, although even these are not always infallible (a researcher should always crosscheck information from one source against other sources to be sure). Wikipedia is

NOT an acceptable source when constructing a scientific paper, since anyone can post information on Wikipedia. It is however acceptable to use popular search engines like Yahoo or Google. Instead of using the regular Google search engine, your best strategy for finding reliable information on the web may be to use Google Scholar. **Google Scholar** is a way to use the Google search engine to specifically search *scholarly and scientific* literature. (www.scholar.google.com). A search utilizing Google Scholar is more likely to turn up articles that have been peer-reviewed.

Peer-Reviewed Articles

When searching for sources, be sure to make the distinction between a peer-reviewed article and a non-peer reviewed article. A peer-reviewed article is an article that has been written by someone with credentials in the field and the article was reviewed by peers in the discipline before the article was accepted for publication. In other words, if looking for environmental science information, the author of a peer-reviewed article is more likely to be an authority in the field of environmental science. Other authorities in the field of environmental science read, review, and approve of peer-reviewed articles before those articles can be published in scholarly journals like *Ecology* or *Environmental Ethics*. A non-peer-reviewed article is one that is more likely to appear in a magazine as opposed to a journal. Always remember that while journals are written to educate, magazines are written primarily to entertain. The articles in a magazine therefore have not had to undergo as much scrutiny as journal articles and may not be as reliable. Also remember that while experts and scholars submit articles to journals, most anyone can write a magazine article, regardless of credentials.

Procedure:

Take notes and listen carefully as the library instructor or laboratory instructor gives a presentation on using the library. You will be tested on this material later.
- Finding Books
- Finding Journals
- Using the World Wide Web

On the Environmental Science Library Research Sheet that follows, answer properly questions 1, 2, and 3. During the semester, you may be required to write *at least one* formal laboratory report. Laboratory reports often require outside information from several reputable sources. The goal of this library exercise is in part to give you a head start on finding the sources you will need should your instructor require you to write a formal laboratory report.

References:

This exercise was developed in part by the UTC library staff. Similar to library exercises found in a variety of sources including:

Cox, G. (2001). *Laboratory manual of general ecology*. Dubuque, Iowa: Wm. C. Brown Publishers.

IZ Quotes. (2014). Luigi Barzini. Retrieved from http://izquotes.com/quote/317276.

IZ Quotes. (2014). James Boswell. Retrieved from http://izquotes.com/quote/317277.

INFORMATION LITERACY DATA SHEET

1. Find 3 **peer-reviewed** articles on cultural eutrophication or on some other assigned topic. Make sure that the articles are available either in the library or full-text online. In the space below, document where you found the article, e.g. database name. If it is available in the library, note its location, e.g. Call Number. Cite one of these articles in the CSE citation style.

2. Find 3 additional articles from other sources (e.g. books, other journals, magazines), and cite them as above. (Again, make sure that the articles are available in the library or full-text online.)

3. Using an Internet Search engine such as Google or Yahoo, search for information on one of the following water quality parameters (or on some other assigned topic): **pH, dissolved oxygen, conductivity, temperature, and turbidity**. Select one website that you trust and one website whose information you don't trust. Give at least 3 reasons for each website you chose.

Basic Statistics

Student Learning Outcomes:

1. To learn the basics of a spreadsheet program.
2. To calculate descriptive statistics by hand and by using Excel.
3. To analyze bivariate data sets.
4. To analyze groups of data using the student's t-test.

Discussion:

The field of statistics is concerned with methods of organizing, summarizing, and interpreting data. The word "data" means information and scientific data are generally numeric (***quantitative***). In the environmental sciences, many scientists collect quantitative data so that statistics can be used to help prove or disprove a hypothesis. There is a high degree of variability within sets of data. As a result, the statistical analysis of data can help describe variability and can be used to compare values in different sets of data. In this laboratory exercise, you will use a spreadsheet computer program to organize, summarize, and interpret data sets.

Descriptive Statistics are measures of the location and spread of data. There are three measurements that indicate the center of distribution in data sets: **(1) Mean (M):** the arithmetic average of the data values, $M_x = (\sum X)/N$) where N equals the number of values in the data set, **(2) Median (Md):** the midpoint of data values when arranged in increasing order, **(3) Mode (Mo):** the value that occurs most frequently. There are also three measurements that indicate the spread in data. These are descriptions of data variability: **(4) Range (R):** the maximum value - the minimum value, **(5) Variance (s^2):** the average squared difference of sampled values from the mean, and **(6) Standard Deviation (s):** the square root of the variance.

Sample calculations:

Mean (M):	Median (Md)	
25	40	
19	37	
40	26	
37	**25**	Md = **25**
26	20	
20	19	
19	19	
$\sum X = 186$	When values are arranged in order from highest to lowest, the median is the middle data point.	

$M = (\sum X/N) = 186/7 = $ **26.6**

Median (Md)

40
37
26   Md = **25.5**
25
20
19
19

When there is an even number of values, take the two middle values and compute the average.

Mode (Mo)

40
37
26
25
20
19   Mo = **19**
19

The mode equals 19 because this value occurs more frequently than any other value.

Mode (Mo)

40
40
26
25
20
19
19

Here there are two modes (**19 and 40**) so the distribution is *bimodal*.

Range (R)

Classroom A scores
160
130
100
  70
  40

R = high score – low score
R = 160 – 40 = **120**

Variance (s^2) = $\dfrac{\sum (X-M_x)^2}{N-1}$

1. First, compute the mean.
2. Subtract the mean from each value $(X - M_x)$
3. Square the values obtained in step 2, $(X - M_x)^2$
4. Sum the squared values, $(X - M_x)^2$
5. Divide by $(N – 1)$. "N" is the number of values.

Values	$X - M_x$	$(X - M_x)$
160	160-100= 60	$(60)^2 = 3600$
130	130-100= 30	$(30)^2 = 900$
100	100-100= 0	$(0)^2 = 0$
70	70-100=-30	$(30)^2 = 900$
40	40-100=-60	$(60)^2 = 3600$

$\sum X = 500$ $\sum(X - M_x)^2 = 9000$
$M_x = 500/5 = 100$

$s^2 = \dfrac{\sum (X-M_x)^2}{N-1} = \dfrac{9000}{5-1} = \textbf{2250}$

Standard Deviation (s) = square root of 2250 = **47.4**

Bivariate Studies involve data sets with information about two different variables. When a data set includes information about two different variables, it is useful to know if there is a relationship between them. For example, if you have data on air temperature AND humidity, it would be important to know if there is a relationship between the two variables and what kind of relationship exists.

Scatter Plots - Plots of the data, with one variable on the x-axis and the other variable on the y-axis, can give us an idea of a potential relationship. If the plot shows that low values of one variable tend to occur with low values of the second variable, a positive linear relationship is shown (Figure 1). If the plot shows that high values of one variable tend to occur in conjunction with low values of another variable, a negative linear relationship is shown (Figure 2). There may be cases in which there is no relationship, or the relationship is not linear (Figure 3).

Figure 1. Positive Linear Relationship

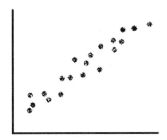

Figure 2. Negative Linear Relationship

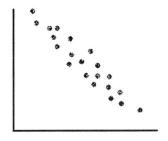

Figure 3. No Relationship

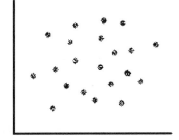

Correlations - Scatter plots showing relationships in picture form are useful, but we often need a more numerical way of expressing the strength or weakness of a relationship between two variables. A correlation coefficient (r) is a quantitative measure of relationship. Correlation coefficient values range from -1.00 to +1.00, representing perfect negative and perfect positive correlations, respectively. A coefficient of 0.00 represents no linear relationship at all.

One of the most commonly used of all statistical tests is the **student's t-test.** The purpose of the t-test is to determine whether the means of two groups of data differ to a statistically significant degree. The test comes in two versions, the paired t-test and the unpaired t-test. A **paired t-test** is used when each data point in one group corresponds to a matching data point in the other group. A typical example in medical research is the same group of patients before and after a treatment. An **unpaired t-test** is used when the groups under study are distinct. In a t-test, if samples are large (>100 data points), the "unequal variances" type of test should be done. For small and moderate sample sizes, the "equal variances" type of t-test provides an exact test of the equality of the two population means (Ford, 2002).

Excel output for a paired t-test (equal variances) is displayed in Figure 4.

Figure 4. t-Test: Two-Sample Assuming Equal Variances

	Variable 1	Variable 2
Mean	66.66667	85
Variance	187.5	150
Observations	9	9
Pooled Variance	168.75	
Hypothesized Mean Difference	0	
Df	16	
t Stat	-2.99382	
P(T<=t) one-tail	0.004295	
t Critical one-tail	1.745884	
P(T<=t) two-tail	0.008589	
t Critical two-tail	2.119905	

Look at the P values (one-tail). If that number is less than 0.05, then the means are significantly different. In this case, the groups are significantly different from each other with a p value of 0.004.

References:

Ford, D. 2002. *Introduction to Environmental Problems II.* Kendall Hunt Publishing: USA.

Quote Garden. (2014). M.J. Maroney. Retrieved from http://www.quotegarden.com/statistics.html.

Quote Garden. (2014). W.A. Wallis. Retrieved from http://www.quotegarden.com/statistics.html.

STATISTICS DATA SHEET

Name: _____

Part I: Calculate (by hand) mean, median, mode, range, variance, and standard deviation for the following data set. Suppose you were interested in the wing lengths of a species of wetland songbird known as the Red-Winged Blackbird. By picking random map locations in Polk County, Tennessee, you captured 30 adult male Red-Winged Blackbirds. You measured the wing length (in mm; often called the **wing chord** [WC] for each of these birds and then released it. The following are the data you collected.

96 99 99 89 102 100 79 99 99 99 89 89 96 96 107 106 99 94 79 89 99 96 95 99 100 100 94 103 95 102

Part II: Now use **Excel (2010)** to calculate mean, median, mode, range, variance, and standard deviation for the same data set. Click on *File* and then click *Options*. Go to *Add-ins*. Go down to *Manage* and select *Excel Add-Ins* and then click *Go*. Select *Analysis ToolPak* and select *Ok*.

Enter the data into the spreadsheet exactly as it appears (one number per cell). Select the *Data Tab* and then select *Data Analysis*. Select *Descriptive Statistics,* click *OK,* and *Input* A1 to A30 with the mouse. Select *Output Range* and click a cell in the spreadsheetwhere you want the data to appear. Select *Summary Statistics* and select *OK*. Drag the columns wider if you need to, to see the words *Standard*

Deviation and *Sample Variance.* Print the results (staple them to this data sheet) and check your hand calculations.

Part III: You will now perform a t-test using Excel. Suppose the entire class takes an extraordinarily difficult environmental science exam (probably not too hard to imagine). The instructor then gives one of his scintillating and endlessly-fascinating lectures and then lets the class take the exam again (that probably *is* hard to imagine). The pre-test and post-test scores are displayed below:

Pre-test scores	Post-test scores
80	90
75	85
60	95
90	75
65	90
45	65
60	70
70	94
55	99
85	80

Using **Excel (2010)**, enter the data in two columns. Under the *Data* tab, select *Data Analysis*, then *t-Test: Two Sample Assuming Equal Variances.* Click *OK.* For variable 1 range, highlight column 1 of the table; for variable 2 range, highlight column 2. Click *OK* and print the results. Answer directly on the print-out whether or not the two groups of scores are significantly different (look at the P value (one-tail). If that number is less than 0.05, then the means are significantly different. Staple the results to the data sheet with your hand calculations.

Part IV: (Bivariate Analysis/Scatter Plots) - You will now perform a scatter plot using the following data set. Please note that this data represents water temperature (in degrees Celsius) and dissolved oxygen (in ppm) for a stream:

Water Temp	DO
22	5
15	8
18	17.5
19	6
20	5.5
18	7
21	5
13	10
16	9
21	5
24	4
12	9.5
19	6.5
15	8.5
18	8
22	4.5
20	6
13	11

Using **Excel (2010)**, enter and highlight the data for both columns of the table. Open the *Insert* tab and click on *Scatter* under *Charts*. Select *Scatter with only Markers*. Use the options under *Chart Tools (Layout)* to name the axes and give the graph a title (use *Chart Title* and *Axis Titles*). Print the chart. Examine the plot and determine the relationship between the two variables (positive linear, negative linear, non-linear, or no relationship). Write the relationship on the plot and staple to the data sheet.

Part V: Get a correlation coefficient for the same data set. **Use Excel (2010)** by selecting a cell (where the correlation coefficient will appear). Click *fx* and type in *Correlation* in the *Search for a Function* box at the top. Hit *Go*. Select *Correl*. For array 1 highlight the numbers in column one. For array 2, highlight the numbers in column two and click *OK*. The correlation coefficient will appear in the designated cell.

   NOTES

Sampling Strategies and Forestry Measurements

> *"Now everybody's sampling."*
> - Missy Elliot

Student Learning Outcomes:

1. To use a compass.
2. To learn different sampling strategies used in the field.
3. To implement the point-quarter method of sampling
4. To perform basic forestry measurements and do basic carbon sequestration calculations.
5. To construct a map showing tree data.

Materials:

Compasses, Surveying Tapes, DBH Tapes, Clinometers

Discussion:

Sampling Methods

Environmental scientists often perform research in the field to obtain qualitative or quantitative data. Qualitative data include descriptive information. Quantitative data include sets of numbers. For detailed research projects, scientists focus on obtaining quantitative data because statistics can be performed in order to evaluate a hypothesis. For biological studies, it is nearly impossible to acquire data on all of the individuals in a given study area. There may be too many individuals or the area may be too big. Therefore, scientists methodically sample a portion of the study area, and there are many sampling strategies used.

How does an environmental scientist mark off appropriate sampling plots (for vegetation and other sedentary organisms)? There are several sampling methods, including the point-quarter method, the line intercept method, and the quadrat method. With the line-intercept method, transect lines are positioned through the area to be studied. Each individual organism or other type of study feature that is intersected by the line or lines is counted or measured (Ford, 2002).

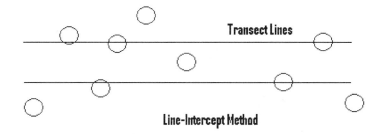

Figure 1. Transect lines in a forested area. *Circles represent trees.*

With the point quarter method, a transect line is run through the study area and sample points are selected along the line. At each sample point, the area is divided into four quadrants. The closest four individuals (one in each quadrant) are measured (Ford, 2002; Figure 2).

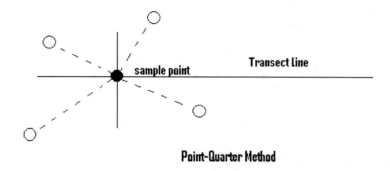

Point-Quarter Method

Figure 2. Point-Quarter method with dotted lines indicating the tree to be sampled in each quadrant.

There is also a quadrat method. A quadrat is an area set aside or laid out on the ground meant to serve as a sample unit. The quadrat can be a square, a rectangle, or a circle. Strip quadrats are long narrow quadrats and are employed when the area to be studied is very narrow). Everything that falls within the quadrat is then counted or measured. Quadrats are useful when measuring plant diversity, especially when sampling herbaceous vegetation or shrubby trees or when the individuals to be studied are extremely abundant (Figure 3). The size of each quadrat depends on the layout of the sample site and the size and dispersal pattern of the individuals to be counted or measured (Ford, 2002).

Q1		Q3		Q5		Q7		Q9	
	Q2		Q4		Q6		Q8		Q10

Figure 3. Quadrats in a sample site.

No matter which sampling regime is chosen, the researcher must use a compass to accurately map the study area. In using a compass, first learn the directions North, South, East, and West. Then learn the parts of the compass shown in Figure 4.

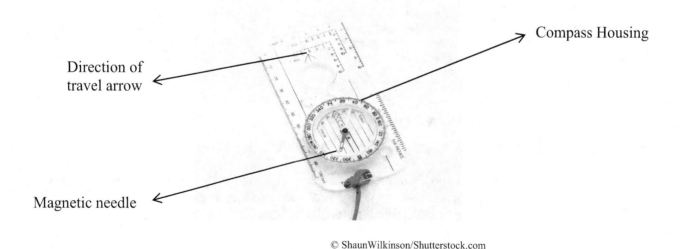

Compass Housing

Direction of
travel arrow

Magnetic needle

© ShaunWilkinson/Shutterstock.com

Figure 4. Parts of the compass.

Always point the compass with the direction of travel-arrows away from you, and into your walking path. The red part of the compass needle is always pointing toward the earth's magnetic north pole (north). The compass housing has a scale from zero to 360 degrees, and the letters N, S, E, W indicate north, south, east, and west. If you know the direction you need to go, simply turn the compass housing to the degree or direction and keep the red needle within the white markers as you walk (Ford, 2002).

Primary Production

Primary production is an important characteristic of ecosystems that affects carbon cycling (Botkin & Keller, 2010). Terrestrial carbon sequestration is the process through which carbon dioxide (CO_2) from the atmosphere is absorbed by trees and other land plants during photosynthesis and stored in biomass. In their capacity to sequester atmospheric CO_2, plants play a critical role in carbon cycling (Botkin & Keller, 2010), both on local and global levels. While accurately calculating the annual **carbon sequestration** of vegetation in a given area could provide information useful in assessing local carbon budgets and needs for offsets, such calculations are inherently difficult due to the complexity of many variables involved. These difficulties include the influence of various plant species and ages, as well as local environmental effects, on carbon sequestration. However, equations developed through previous research on the role of vegetation in carbon cycling do make it possible to estimate the yearly carbon sequestration of a given area.

In this lab you will use the following forestry methods to estimate tree biomass and then use information about the basic biochemistry of a typical tree to determine how much carbon is sequestered by the trees along a transect line.

1. **Tree Inventory**

   Each group of students will evaluate the closest four trees at the 4 assigned points along the transect line using the point quarter method. The first step is to put a flag at the base of each tree that you intend to measure and then measure from the base of that tree to the appropriate point on the transect line. You will also identify each tree by species (don't ask your instructor … it's your job to figure it out). You'll then record the height and DBH (Diameter at Breast Height) using forestry field equipment. You will eventually construct a map of the trees along the transect.

2. **Fresh Biomass**

   Estimates of tree biomass can be calculated from your height and DBH measurements by using **allometric equations**. These equations are commonly used tools by ecologists that relate simple non-destructive measurements – like tree height and diameter – to tree biomass. In many cases, such equations were made of measurements of trees harvested for forestry practices or research studies.

   For this lab, you will use the following allometric equation for estimating the fresh biomass of hardwood (deciduous) tree species in the Southeast developed by scientists for the Georgia Forestry Commission (Clark et al. 1986). It is based on an average of numerous tree species that were studied.

   > For trees with DBH < 11:
   > $W = 0.25D^2H$
   >
   > For trees with DBH ≥ 11:
   > $W = 0.15D^2H$
   >
   > W = aboveground weight of the tree in pounds[*]
   > D = diameter of the tree in inches[*]

H = height of the tree in feet[*]

[*]Although scientists typically use the metric system of measurements, it is common for foresters to employ English units. This may be due to their often close association with logging industry professionals, who use English units.

3. Dry Biomass

According to a University of Nebraska publication (DeWald et al. 2005), the average tree is 72.5% dry biomass (the other 27.5% is water). Use this percentage to estimate the dry biomass of each cemetery tree.

4. Carbon Weight

A USDA Forest Service publication (Birdsey 1992) reports that the average carbon content of a tree is about 50% of its dry biomass. Use this percentage to estimate the carbon weight of each UTC tree.

5. CO_2 Sequestration

Although we can estimate the carbon weight of a tree, a molecule of CO_2 weighs more than a carbon atom. So, you still need to figure out how much CO_2 was sequestered from the atmosphere to construct the biomass of each tree. Consider the atomic weight of a carbon atom and a molecule of CO_2:

$$\text{Atomic weight of C} = 12.001$$
$$\text{Atomic weight of O} = 15.999$$

$$CO_2 = 1\ C + 2\ O$$
$$\text{Atomic weight of } CO_2 = 12.001 + 2(15.999) = 43.999$$
$$\text{Ratio of } CO_2 \text{ to C} = 43.999/12.001 = 3.666$$

In other words, a tree would need to sequester 3.666 g of CO_2 to incorporate 1 g of C in its biomass. Use this ratio to estimate the amount of CO_2 sequestered by each UTC tree.

References:

Botkin, D.B. and Keller, E.A.. (2010). *Environmental science – Earth as a living planet* (3rd ed.). Hoboken, NJ: John Wiley and Sons

Birdsey, R. (1992). Carbon storage and accumulation in United States gorest ecosystems. General Technical Report W0-59. United States Department of Agriculture, Northeastern Forest Experiment Station, Radnor, PA.

Brainyquote. (2014). Missy Elliot. Retrieved from http://www.brainyquote.com/quotes/quotes/m/missyellio279157.html.

Clark A., Saucier J.R., McNab, W.H. (1986). Total-tree weight, stem weight, and volume tables for hardwood species in the Southeast. Research Division, Georgia Forestry Commission.

DeWald, S., Josiah, S., Erdkamp, B. (2005). Heating with wood: producing, harvesting and processing firewood. University of Nebraska – Lincoln Extension, Institute of Agriculture and Natural Resources.

Ford, D. 2002. *Introduction to Environmental Problems II.* Kendall Hunt Publishing: USA.

SAMPLING STRATEGIES AND FORESTRY MEASUREMENTS DATA SHEET

Name: _____

A. Compass with Map

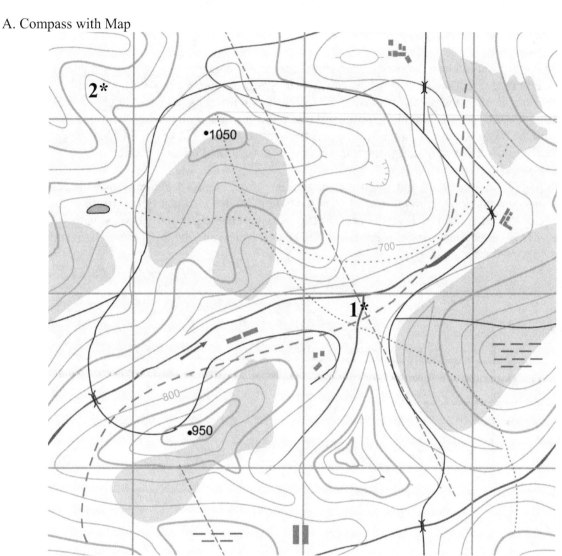

     Imagine that you are traveling from point to point to point on the included map. Locate Point 1 on the map. This will serve as your starting point. Use a ruler to draw a line between Point 1 and Point 2, your destination. Align the compass edge so that the direction of travel-arrows point toward your destination from your origination point. Move the compass housing so that north on the compass matches north on the map (use the meridian line labeled N and S; you may need to extend the meridian lines all across the map). IGNORE THE COMPASS NEEDLE. Read the degrees off the housing where the marker indicates the bearing. Your compass bearing is _____.

     Using these compass bearings, you would be able to walk from one point to another with good accuracy. Your instructor will explain how.

B. Compass in the Field/Placing the Transect

Once in the field, beginning at the point indicated by the instructor, use the tape measure to run a transect line into the sample area. The transect line should be 100 meters long. The assigned compass bearing is _____. This ensures that your transect line will be parallel to all other transects in the study area.

One person should turn the compass housing to the desired degree (the bearing assigned by your instructor). Point the direction of travel arrows away from you, and then rotate your body until the red needle lines up with N on the compass housing. The person with the compass should hold one end of the tape measure. Another person should then walk out into the study area with the tape, with the first person (the one with the compass) directing the second person to move either right or left, in accordance with the compass bearing. Use flags to mark off the transect line at 25m, 50m, 75m, and 100m.

C. At each flag (25m, 50m, 75m, and 100m), locate the four closest trees (deciduous only) in each of the four quadrants (see example below). Place a flag at the base of each tree. Identify the tree to species. Measure the distance to each tree from the central flag. Measure the height of the tree using the clinometer. Measure the diameter at breast height (DBH; at 4 feet) using the DBH tape. Then calculate the CO_2 sequestered for each tree.

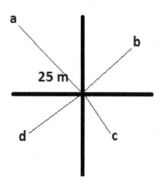

Example

_____

SAMPLE CALCULATIONS:

Use the space below to do a rough approximation of the sampling area, along with any additional calculations.

D. Now construct a more polished map of the area below, specifically of the trees along the two transect lines. You will obviously have to exchange info with the other groups. The map should include the following: the lengths of the transects and the distances of the center flags from the starting points, the compass heading used to mark the transects, the approximate direction of north, the locations of the four trees at each center flag and the distances from the flag to each tree, and the tree data (including the carbon dioxide sequestered). You may use symbols but remember to include a map legend to indicate what the symbols mean.

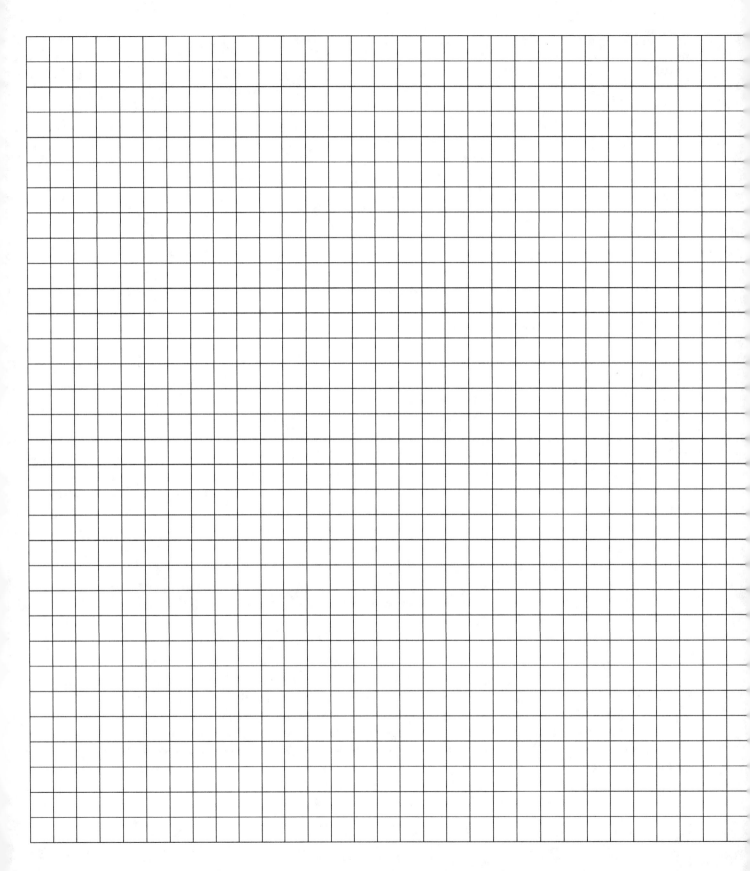

Conservation History: Rachel Carson's Silent Spring

> *"One way to open your eyes is to ask yourself: What if I had never seen this before? What if I knew I would never see it again?"*
>
> *"Only within the moment of time represented by the present century has one species -- man -- acquired significant power to alter the nature of his world."*
>
> – Rachel Carson

Student Learning Outcomes:

1. To become more aware of the dangers of pesticides.
2. To better understand the controversy that surrounded Carson in the 1960's.
3. To better appreciate the impact of the earth-shattering book *Silent Spring*.
4. To become familiar with an important period of time in American conservation history.

Discussion:

Rachel Carson (1907 – 1964) was an influential American marine biologist. Carson was not only an excellent scientist, but a gifted writer as well. She had already had a fair amount of success as a writer when she set out in the 1960s to address the indiscriminate use of pesticides in America. By the time she was done, *Silent Spring*, an incredibly important book, would forever change the way America viewed nature and ecology. It would likewise change the way the world viewed pesticides (which Carson described as "chemicals of broad lethal power" and as "elixirs of death" (Carson, 1962; Goodwin, 2007).

In *Silent Spring*, Rachel Carson challenged not only the chemical companies that manufactured deadly pesticides (most notably DDT), but also the federal government, who wholeheartedly supported and endorsed the use and application of these pesticides on both public and private lands. Carson pointed out in *Silent Spring* that pesticides, improperly applied, had the capacity to kill not only destructive insects but beneficial insects as well (beneficial insects routinely control destructive insects). She also pointed out that other forms of wildlife could be harmed as well, right along with human health and public welfare (Goodwin, 2007).

Simply stated, Carson recognized the true nature of pesticides. She recognized that the ideal pesticide would be narrow spectrum rather than broad spectrum. Narrow spectrum pesticides would kill only the intended target. Broad spectrum pesticides kill the intended target but also other forms of life as well. For that very reason, Carson argued that human beings must be extremely cautious in how they use and apply these lethal chemicals. It should be noted that she did not endorse the outright banning of these chemicals. Carson simply believed that pesticides should be wielded in a more careful manner (given their capacity for destruction) (Goodwin, 2007).

References:

Carson, R. 1962. *Silent Spring*. Boston: Houghton-Mifflin.

Goodwin, N. (Producer). (2007). Rachel Carson's Silent Spring [Motion picture]. United States: PBS Video (The American Experience).

Thinkexist. (2014). Rachel Carson. Retrieved from
http://thinkexist.com/quotation/only_within_the_moment_of_time_represented_by_the/339239.html

RACHEL CARSON'S SILENT SPRING DATA SHEET

Name: _____

Watch the Rachel Carson videos at the following web locations and then answer the questions below to the best of your ability.

http://www.pbslearningmedia.org/resource/envh10.sci.life.eco.silentspring/rachel-carsons-silent-spring/
http://video.pbs.org/video/1442629512/

1) What was Carson's "dire warning?"

2) What do you think the name *Silent Spring* means?

3) Rachel Carson often spoke about the balance of nature. Explain what is meant by the term "the balance of nature."

4) What made this issue so controversial in the 1960s? Why do you think so many people were so threatened by the basic message of *Silent Spring*?

5) Are there parallels between this issue and any modern day issues?

Human Population Growth and Survivorship

"Can you think of any problem in any area of human endeavor on any scale, from microscopic to global, whose long-term solution is in any demonstrable way aided, assisted, or advanced by further increases in population, locally, nationally, or globally?" - Albert Allen Bartlett

"Instead of controlling the environment for the benefit of the population, maybe we should control the population to ensure the survival of our environment." - Sir David Attenborough

Student Learning Outcomes:

1. To understand the effects of the human population explosion.
2. To compare and contrast past and present human mortality.
3. To calculate annual human growth rate and human population doubling time.
4. To calculate and graph survivorship for the U.S population in the 1800's and today.

Materials:

A local cemetery, local obituary pages, a calculator.

Discussion:

The world's human population increased slowly until about 1650. At that time, there were about 500 million people on Earth. In the next 200 years, the population doubled to 1 billion and then it doubled two more times over the next 150 years (Botkin & Keller, 2010). The human population has obviously increased drastically increased over the last 50 years as shown in Figure 1. Currently, there are over 7 billion people on Earth.

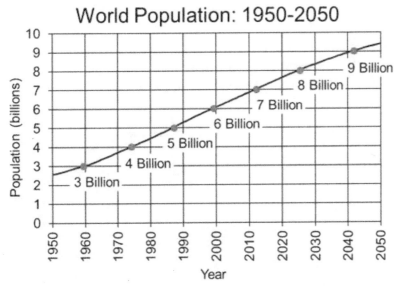

Source: U.S. Census Bureau, International Data Base, July 2015 Update.

Figure 1. World population 1950 through 2050 (est.).

Not only are there now more people on Earth, those people are living longer. The average life span in 1796 was 24 years. By 1896, the average life span had doubled to 48 years. Today, life expectancy in the United States is 78.8 years, with the world average for life expectancy is 71 as compared to the 1948 world average for life expectancy of 48 years (World Health Organization, 2016). This longer life span can of course be attributed to increased access to better nutrition, better sanitation, and better health care (Botkin & Keller, 2010). **More people surviving longer produce more potential parents, which compounds human population growth in the future, straining resources and amplifying pollution.**

Today, we will specifically study the human population by looking at survivorship rates. We will graph survivorship curves for individuals living in the United States in the 1800's and in the 21st century. Survivorship curves are usually one of three general types: Type 1, Type 2, or Type 3.

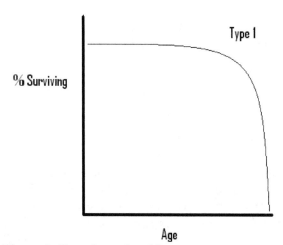

Figure 1. Type 1 survivoship curve.

In the Type 1 survivorship curve, most members of the population die when they reach old age. In the Type 2 survivorship curve (Figure 2, there is an equal amount of death at all ages. In other words, age does not seem to be a significant factor in determining when an individual member of a population will die.

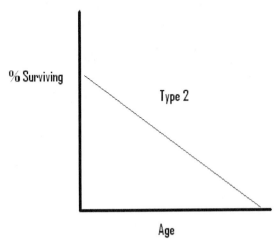

Figure 2. Type 2 survivorship curve.

Finally, in the Type 3 survivorship curve, most individuals exhibit a high incidence of juvenile mortality and die while still very young.

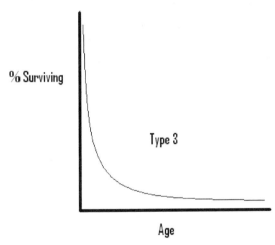

Figure 3. Type 3 survivorship curve.

When constructing survivorship curves, age is plotted on the x-axis and the percentage of people surviving to that age is plotted on the y-axis. As age increases along the x-axis, we will see how people surviving to that age respond and can make some distinctions between human mortality and survivorship in the past versus human mortality and survivorship in modern times. When studying human population growth, it is helpful to calculate annual growth rate and population doubling time for a given country:

- *Annual Growth Rate* = **(Birthrate – Death Rate)/10**
- *Population Doubling Time in Years* = **70/Annual Growth Rate**

Consider the following example: according to the U.S. government's World Fact Book (2013), there are 18.61 births for every 1000 individuals in Mexico. Conversely, for every 1000 individuals in Mexico, there are 4.94 deaths. The annual growth rate of Mexico can be calculated using these values.

(18.61 births – 4.97 deaths)/10 = 1.36 %

An annual growth rate of 1.36 % means that Mexico's population was growing by 1.36 % as of 2013 (excluding emigration and immigration). **Emigration** is the movement of people out of a country, while *immigration* is movement of people into a country,

At this rate, it can be calculated how long it will take for Mexico's population (*116,220,947*) to double:

70/1.36 = 51.5 years

By 2065, at the current rate, Mexico's population will be *232,441,894*.

References:

Botkin, D.B. and Keller, E.A.. (2010). *Environmental Science – Earth as a living planet*. Hoboken, NJ: John Wiley and Sons

Central Intelligence Agency. (2013). *The world factbook*. Washington, D.C.:

IZ Quotes. (2014). A.A. Bartlett, Retrieved from http://izquotes.com/quote/368373.

Metafilter. (2014). David Attenborough. Retrieved from http://www.metafilter.com/108752/Instead-of-controlling-the-environment-for-the-benefit-of-the-population-maybe-we-should-control-the-population-to-ensure-the-survival-of-our-environment-Sir-David-Attenborough

World Health Organization. (2016). Global health observatory data. Available at http://www.who.int/gho/mortality_burden_disease/life_tables/situation_trends/en/.

 NOTES

Energy Conservation, Air Pollution, and *Kilowatt Ours*

> *"Energy is an eternal delight, and he who desires, but acts not, breeds pestilence."*
> – William Blake

Student Learning Outcomes:

1. To know the sourcesof our energy.
2. To understand the connection between the burning of coal and air quality.
2. To identify simple ways to conserve energy in our daily lives.
3. To learn about green energy.
4. To become aware of the Tennessee Valley Authority's Green Power Switch.

Discussion:

The majority of our electricity in the United States is generated through the burning of coal (Power Scorecard, 2014). Coal is burned in coal-fired power plants to produce great amounts of heat and this heat is used to turn water into steam. Steam turns a turbine which spins a generator producing electricity. For example, the Tennessee Valley Authority's Kingston Power Plant burns 14,000 tons of coal each day and powers 700,000 homes a year (Tennessee Valley Authority [TVA], n.d.).

Unfortunately, the harvesting and burning of coal to produce electricity can adversely impact the environment and human health (Environmental Protection Agency [EPA], 2013). Coal is mined from the earth using techniques such as surface mining, underground mining, and mountaintop removal. Mining of any type can generate air pollution, water pollution, and noise. Mountaintop removal is especially controversial because the method involves using explosives to destroy the upper topography of the mountain, providing easy access to the seams of coal contained therein (EPA, 2010). The burning of coal has been linked to acid rain, air pollution, water pollution, greenhouse gas emissions, smog formation, reduced visibility, and increased rates of asthma. When one considers the adverse consequences of extracting and burning coal, along with the fact that coal is a nonrenewable resource, many scientists believe that now is the time to expand the use of alternative forms of energy.

What about *nuclear energy*? TVA's nuclear plants produce about 30% its power supply (TVA, 2013).The way nuclear power works is that uranium pellets are placed in rods and the rods are bombarded with neutrons. Nuclear fission takes place and heat is released. The heat converts water to steam and the steam again turns a turbine which spins a generator that produces electricity (TVA, 2013). Admittedly, nuclear power is much cleaner than the burning of coal. The problems associated with nuclear power include radioactive waste and potential nuclear accidents and meltdowns at nuclear power plants.

More and more, our society is slowly moving toward alternative sources of energy. Instead of relying on energy from coal-fired power plants, people in the Southeast and across the country are utilizing clean and renewable sources of energy. *Green energy* is energy produced by the sun, the wind, and other sources (TVA, 2014). These sources are considered renewable meaning that we will never run out of solar power and wind power. These types of energy produce less waste and less pollution than the sources of energy that we have traditionally relied upon.

Because of the environmental effects of current energy sources, it is important that people practice *energy conservation*. There are simply ways to conserve energy. For example, simply turning off the lights when you leave a room. Shutting down your computer at night. Unplugging your cell-phone charger when it's not in use. Turning off your television when you leave a room. Replacing your incandescent light bulbs with the more energy efficient, longer lasting bulbs. There are so many simple ways to conserve energy! If we use less energy, then less coal will be burned in power plants, reducing the associated adverse environmental impacts. Increased energy efficiency will also save you and your family money.

Your household may want to consider using the savings from your newfound energy efficiency to purchase green power from a TVA program called the ***Green Power Switch***. The Green Power Switch allows residential customers to purchase blocks of Green Power. Each block costs $4, is added to your monthly bill, and ensures that 150 kilowatt-hours of electricity is generated by a renewable source of power (TVA, 2014). Four or more dollars per month, as pointed out in today's video, is a small price to pay for the peace of mind that comes with being more energy-conscious and more environmentally-friendly.

CHECKLIST OF SİMPLE WAYS TO SAVE ENERGY
□ Take shorter showers
□ Line dry clothes, when possible
□ Set thermostat at 68 degrees or lower in the winter
□ Wash clothes in cool or cold water, when possible
□ Unplug cell phone charger when not in use
□ Open dishwasher door and let dishes air dry
□ Lower the temperature on the water heater
□ Wear extra clothes when cold, instead of turning up thermostat
□ Close air vents in unused rooms
□ Switch to/use compact fluorescent bulbs

References:

Brainyquote. (2014). William Blake. Retrieved from http://www.brainyquote.com/quotes/quotes/w/williambla150122.html

Environmental Protection Agency [EPA]. (2013). Coal. Retrieved from http://www.epa.gov/cleanenergy/energy-and-you/affect/coal.html.

Environmental Protection Agency [EPA]. (2010). EPA science on mountaintop mining. Retrieved from http://www.epa.gov/sciencematters/august2010/mountaintop.htm.

Kilowatt Ours. (2008). *Kilowatt Ours – A Plan to Re-Energize America* (Companion Curriculum). Retrieved from www.kilowattours.org/educators.

PowerScorecard. (2014). Electricity from coal. Retrieved from http://www.powerscorecard.org/tech_detail.cfm?resource_id=2.

Tennessee Valley Authority [TVA]. (n.d.). Coal-fired power plant. Retrieved from http://www.tva.gov/power/coalart.htm.

Tennessee Valley Authority [TVA]. (2014). Green Power Switch. Retrieved from http://www.tva.com/greenpowerswitch/index.htm.

Tennessee Valley Authority [TVA]. (2013). Nuclear energy. Retrieved from http://www.tva.gov/power/nuclear/index.htm.

ENERGY CONSERVATION, AIR POLLUTION AND *KILOWATT OURS* DATA SHEET

Name: _____

1) Why do you think Americans today use so much electricity?

2) Think about "green power." Do you think nuclear power is green power? Why or why not?

3) Why do you think Jeff Barrie called his film "Kilowatt **Ours**?"

4) Coal accounts for 52% of electricity generated in the U.S. The average home uses 920 kWh a month. It takes 2.25 lbs. of coal to make 1 kWh of electricity. How much coal is required to generate enough electricity for one single day in an average home?

5) Can you pick five actions that you could commit to in order to help save energy? If so, what are the five? See the checklist on the previous page for inspiration.

Energy Resources

Student Learning Outcomes

1.   To learn about different sources of renewable and nonrenewable energy.
2.   To practice making energy efficiency measurements / doing basic energy calculations.
3.   To learn about alternative energies.

Discussion:

Electricity is used in almost every aspect of our lives. To generate the electricity we need, stored potential energy must be converted to useable kinetic energy (U.S. Energy Information Administration [EIA], 2013). The United States is the largest consumer of electricity in the world, and most of our energy comes from ***non-renewable*** fossil fuels such as oil, gas, and coal. These resources exist in a fixed amount. With the rapid use of these fuels, nonrenewable resources will eventually be depleted (U.S. EIA, 2013). Nearly 90% of the electricity used in the United States is produced through ***steam generation***. In this process, a fuel (fossil, nuclear) with potential chemical energy is changed to kinetic heat energy. Kinetic heat energy is used to heat water for steam, and steam turns a turbine that operates a generator, producing electrical energy (U.S. EIA).

In general, ***fossil fuels*** have formed from dead plant and animal matter that has endured intense heat and pressure over millions of years (U.S. EIA, 2013). These fuel sources are widely used because they are abundant, easy to transport, and relatively inexpensive. ***Coal*** is a rock-like fossil fuel consisting mostly of carbon, with small amounts of water, sulfur, and radioactive materials. Coal is removed from the earth by surface or underground mining, is processed through crushing and washing, and is eventually shipped to power plants (U.S. EIA, 2013). Coal is burned to produce steam that turns turbines to generate electricity. Some air pollution is released as coal is burned including sulfur dioxide, nitrogen oxides, particulates, carbon dioxide, mercury and other heavy metals, and ash. Devices in smokestacks such as scrubbers remove some of the air pollution (U.S. EIA, 2013) .

Petroleum is a gooey liquid of mostly carbon with small amounts of sulfur, nitrogen, and oxygen (U.S. EIA, 2013). Crude oil (oil as it comes out of the ground) is pumped out of rock formations by wells or oilrigs. At refineries, crude oil is separated into different types of products such as diesel fuel, heating oil, gasoline, and asphalt (U.S. EIA, 2013). In the United States, most oil is used in the form of gasoline.

Natural gas is a gaseous mixture of mostly methane with small amounts of ethane, propane, butane, and hydrogen sulfide (U.S. EIA, 2013). Once a deposit of natural gas is located, it is tapped and pumped out of the deposit. Through processing, impurities are removed, propane and butane are separated and used for heating and cooking fuel, and methane is separated and used to heat buildings and to generate electricity (Environmental Protection Agency [EPA], 2013).

Renewable energy sources are essentially inexhaustible on a human time scale and include solar, wind, and flowing water energy. During every 24-hour day, more than 30,000 times as much energy mankind uses in one day will reach the earth as radiant energy. To use solar energy, we must take this diffuse energy source and concentrate it into a form that can be used for heat and electricity (U.S. EIA, 2013).

Buildings and water can be heated by the sun in two ways: passive and active solar heating. ***Passive solar heating*** systems generally involve building designs that take advantage of solar energy for heating and do not require mechanical power (U.S. Department of Energy [DOE], 2013). The simplest form of passive heating is to allow sunlight to enter the home through south-facing windows. ***Active solar heating systems*** involve pumps and fans that circulate heated air or fluid from roof-mounted solar collectors to radiators or water heaters in the home (U.S. EIA, 2013).

Photovoltaic cells (PV) can be used to generate electricity from solar energy. PV cells are made up of thin silicon wafers, and many cells are wired together to form a panel (U.S. EIA, 2013). A solar cell converts solar energy to DC electricity, which can then be stored in batteries or used directly by conversion to an AC current (U.S. EIA, 2013).

Incandescent Light Bulb Experiment

Types of energy used should be evaluated in regard to our quality of life and environmental effects. So, what is our best energy option? Most scientists believe we should focus on these two goals:
1) Improving energy efficiency.
2) Using more renewable energy and less nonrenewable energy.

One way to use more of the energy we produce is through cogeneration – the use of waste heat for other purposes. In this laboratory exercise, you will examine a form of waste heat. Perhaps the most common household example of wasted energy is the heat given off by traditional incandescent light bulbs (U.S. DOE, 2012). This type of bulb produces light by allowing a flow of electricity through a filament, heating the filament until it glows. You can feel the heat if you put your hand near a bulb that has been burning for a while

Energy-efficient light bulbs such as energy saving incandescents, CFLs and LEDs use 25% to 80% less energy and also produces less waste heat (U.S. DOE, 2012).. In today's activity, you will measure the heat given off by a small incandescent bulb using a simple calorimeter.

Figure 1. Set-up of a simple calorimeter (TVA, 1992)

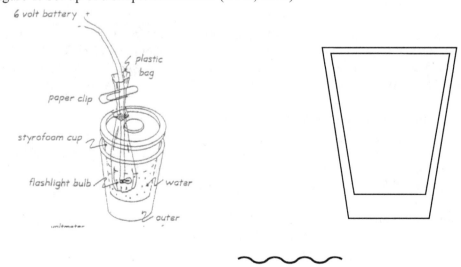

References:

Environmental Protection Agency [EPA]. (2013). Natural gas. Retrieved from http://www.epa.gov/cleanenergy/energy-and-you/affect/natural-gas.html.

Tennessee Valley Authority [TVA]. (1992). Energy Sourcebook. Retrieved from http://www.tvakids.com/teachers/pdf/elementary_sourcebook.pdf.

Thinkexist.com. (2014). Ralph Marston. Retrieved from http://thinkexist.com/quotation/you-ve_done_it_before_and_you_can_do_it_now-see/223893.html.

U.S. Energy Information Administration [EIA]. (2013). Energy explained. Retrieved from http://www.eia.gov/energyexplained/index.cfm.

U.S. Department of Energy [DOE]. (2012). New lighting standards being in 2012. Retrieved from http://energy.gov/energysaver/articles/new-lighting-standards-begin-2012.

U.S. Department of Energy [DOE]. (2013). Passive solar home design. Retrieved from http://energy.gov/energysaver/articles/passive-solar-home-design.

ENERGY RESOURCES DATA SHEET

Name: _____

Procedure and Initial Measurements - Working in groups, follow these steps:

- Put one Styrofoam cup inside another of the same size.
- Measure out 150 ml of water with a graduated cylinder and fill the inside cup, recording the water volume and mass in the table below (1 ml of water has a mass of 1 gram).
- Attach one piece of wire to one of the light bulb filaments. To do this, pull the exposed wire through the loop of the bulb filament and twist the wire together. Repeat this procedure with the second piece of wire and attach to the other filament of the bulb.
- Check to make sure connections are made by attaching the two free ends of the wires (with alligator clips) to separate battery posts. The bulb will light up if the circuit is complete. Detach the wires from the battery while completing the assembly of the rest of the apparatus.
- Place the bulb in the bottom of a plastic sandwich bag. Make sure the wires remain attached to the bulb while pressing the bag from the bottom up to exclude most of the air. Use a paperclip to secure the top of the bag around the wire leads.
- Pull the bottom of the bag through the slit in the cup lid and set aside.
- Take an initial temperature reading of the water and record. Secure the lid on the cup so that the part of the bag with the bulb is resting in the water.
- Connect the alligator clips to the battery (the light should be on). Wait thirty minutes. Then detach the battery, open the calorimeter, and take the water temperature and record.
-

Volume of Water (mL)	Mass of Water (g)	Initial Temp of Water (C)	Final Temp of Water (C)	Temperature Change (C)	Specific Heat of Water
					1 Cal / (g C)

Heat Energy (calories) = mass x specific heat x temperature change = _____ calories

Heat Energy (joules) = calories x 4.2 = _____ joules

B. Calculate Power and Energy from Data Given by the Instructor

Voltage (V)	Current (I) Amps	Power (watts) = V x I	Time (seconds)	Total Energy (joules) = power x time

C. Calculate Light Energy and Percent Efficiency

Total energy (joules) produced by battery	
Heat energy (joules) produced by battery	
Light energy = Total energy – heat energy	
% efficiency = **(light energy/total energy) x 100**	

Biodiversity and the Biodiversity Crisis

"We should preserve every scrap of biodiversity as priceless while we learn to use it and come to understand what it means to humanity." – E.O. Wilson (Biodiversity Conservancy International, 2012)

"What is man without the beasts? If all the beasts were gone, man would die from a great loneliness of spirit. For whatever happens to the beasts, soon happens to man. All things are connected." – Chief Seattle

Student Learning Outcomes:

1. To learn basic information about the biodiversity crisis.
2. To consider the variety of life on earth by examining a variety of birds common to the area.
3. To gain practical experience identifying birds with a field guide
4. To understand the importance of bird conservation and make a contribution to a real-life bird monitoring project.

Materials:

Saving Species DVD, field guides, clipboards, binoculars (optional)

Discussion:

Biological diversity, or biodiversity for short, is perhaps best defined as the variety of life on earth. Three different levels of biodiversity exist. The three levels are diversity of genetics, diversity of ecosystems, and diversity of species. **Environmental scientists and conservation biologists are very concerned with preserving and protecting all three levels of biodiversity.** As for species diversity, some estimate that there are anywhere from 5 to 30 million species in existence. Other estimates place the total number of species on Earth at around 100 million. Scientists have documented approximately 2 million species (mostly insects). This means that there are literally millions of species yet to be discovered (Reece, Taylor, Simon, & Dickey, 2015; Biodiversity Conservancy International, 2014))!

Extinction is the elimination of every individual of a particular species. The term 'background extinction' refers to a normal, continual, ongoing, low-level elimination of species. A 'mass extinction,' on the other hand, is the extinction of many species in a relatively short period of time. The dinosaurs suffered a mass extinction at the end of the Mesozoic era (245 mya to 65 mya), probably due in large part to an asteroid impact that kicked up dust and blocked out the sun, triggering a major climate change. Unfortunately, a mass extinction of species is currently under way. This mass extinction of species, however, is very different than the Cretaceous extinction of the dinosaurs. Unlike the Cretaceous extinction of the dinosaurs and every other mass extinction in the history of our planet, this current mass extinction is caused by human activity. The problem is this: **species currently go extinct at a much faster rate than they did just a few hundred years ago.** The human population of the Earth is constantly increasing and as the human population increases, biodiversity declines (Reece, Taylor, Simon, & Dickey, 2015).

The primary threat to biodiversity is habitat alteration. When humans log a forest or build a road through a wilderness or fill in a wetland and construct a parking lot, they are altering and degrading habitat and this reduces species diversity. The second major threat to biodiversity is the introduction of exotic species. Humans intentionally and unintentionally move species from one ecosystem or geographic location to another. An invasive species is likely to have no natural predators or parasites in its new environment to hold its populations in check. With no natural enemies, invasive species will greatly multiply and outcompete native species, occasionally to extinction, resulting in an overall decrease in species diversity. The invasive may even prey directly on the native or somehow otherwise alter community structure to the point that the native cannot survive. The third major threat to biodiversity is overexploitation. Overexploitation includes over fishing, over hunting, and over harvesting. For instance, many commercially desirable species of fish are depleted in many of the world's fisheries due to overexploitation (Reece, Taylor, Simon, & Dickey, 2015).

Focus on Birds

Birds belong to the class Aves. They are vertebrates, meaning that they possess true backbones. Birds of course are covered in feathers and have wings. Birds use their feathered tails for balance and for flight. The bodies of birds are generally very light. This is in part what makes flight possible. The light bodyweight of birds is achieved through 'hollow' bones. The bones of birds, in contrast to the bones of other animals, are empty except for air and thin struts that give the bones strength and structural integrity. The bills of birds are also very light when compared with the heavy toothed-jaws of mammals and reptiles and this also contributes to a lighter overall bodyweight. To maintain a light bodyweight, the digestive systems of birds are extremely efficient. This ensures that birds are not unnecessarily weighted down by the presence of heavy undigested food. Birds also maintain a light bodyweight by laying eggs and reproducing externally. Reproduction that utilizes external eggs (as opposed to the internal development of young) reduces the amount of time that female birds are burdened with the weight of their offspring. In addition to light bodies, birds have also evolved four-chambered hearts, compact lungs, muscular gizzards, large brains, superior coordination and balance, and excellent vision and hearing.

Birds are fascinating creatures and their ranks are extremely diverse. The diversity of birds has come about in part because the birds have adapted to their specific environments and available food sources. Consider, if nothing else, the wide range of beak adaptations common in birds. Some birds are seed eaters and therefore possess short, thick, heavy bills that are useful for cracking seeds. The birds of prey are meat-eaters and possess sharp, curved bills. Curved bills are ideal for tearing flesh. Other birds often have thin, pointed bills that are perfect for grasping insects. Some birds with grasping bills are even tool users. For example, the woodpecker finch of the Galapagos (one of Darwin's famous finches) grasps the spines of cacti. The woodpecker finch can manipulate a spine and use it to dislodge grubs from trees. It should be noted that there are many other differences among the various bird species in terms of feet, legs, wings, coloration, and behavior (Alsop III, 1997, 2001; Sibly, 2001).

There are approximately 10,000 species of birds. According to Birdlife International (2016), 1 in 8 bird species are in trouble. At least 200 bird species are considered Critically Endangered with many more classified as Threatened. We are of course in the midst of a human-driven mass extinction called the Biodiversity Crisis. **Birds, like many other animals, are extremely vulnerable to habitat destruction and degradation, to the introduction of exotic species, and to overexploitation.** As for overexploitation, birds are especially threatened by commerce and by illegal trading since so many bird species are in high demand as pets and for zoos. Although the exact extinction rate for birds is not known, at least 129 species of birds have become extinct since 1600, with many of these extinctions occurring in the last 150 years. Since 1986 for example, the Hawaiian Crow, the Alagoas curassow, the Guam rail, the Kauai 'O'o, the Kama'o, and the Atitlan grebe have all become extinct. Fortunately, some land trusts have been set up to protect birds and many important areas with large numbers of birds have been identified. The identification of such areas is the first step in habitat preservation and in bird protection. It should be noted that the United States government strives to protect birds through compliance with certain international treaties and through laws such as the Wild Bird Conservation Act and the Endangered Species Act of 1973 (Alsop III, 1997, 2001; Sibly, 2001).

There is in existence in the United States a Bird Banding Program (U.S. Geological Survey, 2015). Trained ornithologists and wildlife biologists with the appropriate permits from the federal Bird Banding Laboratory are allowed to capture birds in mist nets in order to place metal bands on their legs. These metal bands are marked with a banding number. The bander then records the banding number, the species name, the age, the sex, and the place and date of banding. Once a captured specimen is properly banded and all of the important information about the specimen is recorded, the bander releases the bird. The information is entered into a database. If someone later recaptures or finds a banded bird, that person makes a report to the Bird Banding Laboratory. In this way, scientists are able to keep track of bird distribution, bird movement, relative numbers, lifespan, cause of death, etc. Banding provides much of the information so crucial to proper bird management and conservation (U.S. Geological Survey, 2015).

<u>Using a Field Guide, Calculating A Diversity Index, and the Concept of a Community:</u>

In today's lab, using field guides and binoculars, you will identify and keep track of the various birds that you see at a nearby bird feeder. The point of this activity is to encourage you to use your powers of observation and also to get you out of the lab and into nature. You will likewise make some basic ecological calculations for two imaginary bird communities. Today you will calculate a diversity index. It should be noted that a biodiversity index is most helpful and useful when you can compare it with another biodiversity index from another site. In this way, you can compare relative diversity from one site to another and see which of the two communities is more diverse. A *community* is best defined as different populations of different species living in the same place at the same time. The various populations that make up a community share resources. They interact with one another. These populations often depend upon one another to a great extent. Even though we will focus today on birds, other forms of life (including plants and insects) also hold key positions within communities. **It is important to understand that human activity (especially urban development and agriculture) gives way to habitat destruction and sometimes adversely impacts communities and alters community structure. We as human beings should therefore be cautious in the way we manage communities and ecosystems.**

References:

Alsop III, F.J. (2001). *Birds of North America – Eastern region.* DK Publishing, Inc: USA.

Alsop III, F.J. (1997). *All about Tennessee birds.* Birmingham, Alabama:Sweetwater Press.

Biodiversity Conservancy International. (2012). Biodiversity 101. Retrieved from http://biodiversityconservancy.org/biodiversity101.html .

Birdlife International. (2016). Saving Species. Available at http://www.birdlife.org/worldwide/science/saving-species.

Quotations Book. (2014). Chief Seattle. Retrieved from http://quotationsbook.com/quote/2600/.

Reece, J.B., Taylor, M.R., Simon, E.J., Dickey, J.L., and Hogan, K.A. (2015). Campbell Biology: Concepts and Connections. Benjamin Cummings: San Francisco.

Planet Earth. (2007). Saving Species Summary. Retrieved from
http://catalog.youranswerplace.org/(riylqzaouv3t2j55dvrnwt21)/NewCopyListing.aspx?QS0=&QS1=2&
QS2=Documentary%20television%20programs%20Videorecordings&QS3=1&QS4=2&QS5=0&QS6=9
9&QS7=0&QS8=9999&QS9=0&QS10=99&QS11=703531&QS12=1&QS13=1&QS14=538013&QS15
=9781419849367

FeederWatch. (n.d.). Project Feeder Watch Overview. Retrieved from
http://feederwatch.org/about/project-overview/

Schneider, R.L.,Krasny, M.E., and Morreale, S.J.. (2001). *Hands-on herpetology – Exploring ecology and conservation.* Arlington, VA: NSTA Press.

Sibly, D.A. and the National Audubon Society. (2001). *The Sibley guide to bird life and behavior.* New York: Alfred A. Knopf.

U.S. Geological Survey. (2015). The North American Bird Banding Program. Available at
https://www.pwrc.usgs.gov/bbl/.

BIODIVERSITY AND THE BIODIVERSITY CRISIS DATA SHEET

Name: _____

A. Today we will watch a film entitled "Saving Species" from the acclaimed BBC TV series, *Planet Earth*. "Saving Species" is described this way: "Many of the animals featured in *Planet Earth* are endangered so do we face an extinction crisis? "Saving Species" asks the experts if there really is a problem, looks at the reasons behind the declining numbers of particular animals and questions how we choose which species we want to conserve" (Planet Earth Saving Species Summary, 2007).

 TAKE NOTES BELOW:

B. Bird Feeder Observation - List number of individuals of each species in view at one time at the feeder (this ensures you will not count the same bird more than once). Limit your observation time to 30 minutes for the purposes of this exercise. The data your class collects will be submitted to Project Feeder Watch associated with the Cornell Lab of Ornithology. Project Feeder Watch is a "winter-long survey of birds that visit feeders at backyards, nature centers, community areas, and other locales in North America. Feeder Watchers periodically count the birds they see at their feeders from November through early April. Feeder Watch data help scientists track broad scale movements of winter bird populations and long-term trends in bird distribution and abundance." See the following links:

http://www.birds.cornell.edu/pfw/

http://www.birds.cornell.edu/pfw/InstruxandUpdates/inst_video.html

Common Name	*Latinized Scientific Name*	NUMBER of INDIVIDUALS
1.		
2.		
3.		
4.		
5.		
6.		
7.		
8.		
9.		
10.		

C. Shannon Diversity Index Applied to two imaginary Bird Communities - Calculate below the diversity and maximum diversity of the bird community at Feeder A and at Feeder B and write in your answers below. See the explanation of the Shannon Index on the following page and show all of your work. DO THIS IN CLASS TODAY!

D. Jaccard Index Applied to Two Different Bird Feeders - Determine how similar Bird Community A is to Bird Community B. See the explanation of the Jaccard Index on the following page and show all of your work below. DO THIS IN CLASS TODAY!

The Shannon Index:

$$H' = -\Sigma p_i \times \log (p_i)$$

Example: You spot 3 different bird species: A, B, and C. There are 4 individuals in A, 5 individuals in B, and 10 individuals in C. First, add up the total number of specimens.

$$4 + 5 + 10 = 19$$

For Species A:	4/19 = 0.211	this is p_1 for A
For Species B	5/19 = 0.263	this is p_2 for B
For Species C	10/19 = 0.526	this is p_3 for C

Now apply the equation: $p_i \times \log (p_i)$

For Species A:	0.211 X log (0.211) =	-0.143
For Species B:	0.263 X log (0.263) =	- 0.153
For Species C:	0.526 X log (0.526) =	- 0.147

The 'Σ' in the equation is a summation symbol, so we should add these values together. Remember also that there is a negative in front of the summation symbol. Therefore, we should multiply by -1 in order to change the sign.

$$H' = -1 \, (-0.143 + -0.153 + -0.147)$$
$$H' = -1 \, (-0.443) \text{ or } 0.443$$

Now calculate H'_{max} for the community. H'_{max} represents the maximum diversity for the community. In our example, there were 3 total species, so:

$$H'_{max} = (\log 3)$$
$$H'_{max} = 0.477$$

This 0.477 is the maximum diversity a community with exactly 3 bird species can have.

You should then compare H' and H'_{max} and see how close the two numbers are to one another. The closer H' approaches H'_{max}, then the more equally abundant the species in the community and the more diverse the community. It might actually prove helpful to divide H' by H'_{max}.

Jaccard Index:

For two sites A and B, it is possible to determine how similar the two sites are in terms of the species common to both sites. A working knowledge of which species are present and which species are absent at each feeder is crucial. If the coefficient is close to 0, then the communities are not similar. The closer the coefficient is to 1, the greater the similarity between the two communities. Note the following formula.

Coefficient of similarity = $a \div (a + b + c)$

a = number of species common to both sites
b = number of species in Site A, but not in Site B
c = number of species in Site B, but not in Site A

Soil Analysis Part 1: pH, Nutrients, Conductivity

"The nation that destroys its soil destroys itself."
- Franklin D. Roosevelt

Student Learning Outcomes:

1. To understand the importance of healthy soil.
2. To learn about the seven primary functions of soil
3. To learn about soil formation (the five soil-forming factors).
4. To learn about and test soil samples for three basic chemical soil indicators.

Discussion:

Environmental scientists around the globe strive not only to protect and preserve air and water quality, but also to protect and preserve the earth's important soil resources. Soil is more than just lifeless dirt. **Soils are extremely important for every land-based ecosystem and for life on earth in general.** Healthy soils of high quality and good fertility are important because they (1) sustain plant and animal life and support biodiversity, (2) support agriculture, (3) produce, convert, and store important gases, (4) detoxify pollutants, (5) store, regulate, and purify water, (6) cycle nutrients, and (7) support buildings and other man-made structures. Soil is essentially a non-renewable resource since it takes so many years to form. Our soil resources are therefore limited. However, if managed properly, the same soil can be used over and over again. Always remember that soil management affects soil quality. Therefore, as environmental scientists and as good environmental stewards, we should strive to protect and maintain high soil quality and good soil fertility.

Soils support massive amounts of animal life (Cornell University, 2014). Soils contain organisms such as snails, slugs, earthworms, roundworms, crustaceans, millipedes, centipedes, insects, and arachnids. Soils also contain literally billions of microorganisms. Some of these organisms break down organic material and free up nutrients for recycling while other organisms aerate the soil by burrowing through it (Cornell University, 2014). Soils also encourage plant growth and support a wide array of plant life which in turn supports the primary consumers which in turn supports the secondary consumers and so on. Healthy soils are essential for agricultural plant production also. Successful agricultural production is very important because the 7 billion people in the world must be fed.

Soils are likewise important because they produce, convert, and store gases that are essential for the proper functioning of the global environment. For example, microbes in the soil called nitrogen fixers convert atmospheric nitrogen (which plants can't use) to nitrates and ammonium compounds (which plants can use and must have for growth) (Cornell University, 2014). Some soil microbes likewise convert carbon monoxide to carbon dioxide. Soils tend to detoxify (or immobilize) pollutants and the conversion of carbon monoxide to carbon dioxide is a good example. Soils store water as well. This is an especially important function of soil, since plants and crops must have water for good health and for the production of biomass. Soils also regulate and purify water. As water passes through soil, the water is filtered of impurities by various soil processes. The purified water may then seep into groundwater.In addition, soils are important because they provide and recycle essential elements and nutrients. only in smaller amounts.

How is soil formed? Five soil forming factors are generally considered in explaining soil formation (National Estuarine Research Reserve System [NERRS], 2011). These five factors are as follows: parent material, climate, topography, biological factors, and time. ***Parent material*** (which may include rocks or other materials located on site or moved around by water, ice or wind) is broken down by weathering. ***Weathering*** is the formation of mineral matter from parent rock (University of Michigan, 2013). Climate also influences soil formation in that different temperatures and moistures affect weathering and leaching. ***Leaching*** is when water percolates in the ground and dissolves and transports minerals in the ground (University of Michigan, 2013).Topography plays a role in soil formation as well since the slope of a piece of land can affect moisture and temperature which in turn affects erosion and weathering and leaching. Plants and animals (especially microorganisms and burrowing animals) help mix and aerate the soil thereby influencing soil formation. Of course, time is required for soil formation. It actually takes 100-400 years for 1 centimeter of top soil to form, essentially making soil a non-renewable resource.

People harm the soil in a variety of ways. Fertile soil can be eroded away when farmers over cultivate, overgraze, and over plow their fields (University of Michigan, 2013). Erosion can also occur when foresters engage in clear cutting. Their equipment can dislodge the soil. Once the trees are gone, roots no longer hold the soil in place. Erosion can lead to soil compaction and moisture loss which might ultimately lead to desertification. Over watering the land can reduce the amount of oxygen in the soil and choke out vegetation. Sometimes irrigation can be destructive to soil. When irrigation water evaporates it can cause dissolved salts to be pulled up out of the ground. This can cause the upper soil layers to become highly saline and less productive. The overuse of pesticides can also chemically contaminate the soil. Again, since soil is essentially non-renewable, we must use the soil in a sustainable way in order to protect and preserve our soil resources for the future.

Soil Acidity

Soil pH can affect microbial and chemical reactions and the solubility and availability of soil nutrients. pH is best defined as the negative logarithm of the hydronium ion concentration in the soil (Colorado State University Extension, 2011). The hydronium ion is a charged particle formed when a proton (H^+) becomes bound to a water molecule (H_2O). The hydronium ion is represented this way:

$$H_3O^+$$

$$pH = - \log [H_3O^+]$$

The ***pH*** and the presence and amount of the hydronium ion are directly related to acidity. The pH scale runs from 0 to 14. The strong acids are from 0 to 2. The moderate acids are from 2 to 5. A pH of 7 is neutral. Anything above 7 is considered basic. The strongest bases are around 13 or 14 (Colorado State University Extension, 2011). See below:

0 ◀ stronger acids   weaker acids   7   weaker bases   stronger bases ▶ 14

neutral

Why be concerned with the pH of a soil? In part, we are concerned about the pH of a soil because *certain plants have very specific pH requirements*. For example, vegetables, grains, trees, and grasses grow well in slightly acidic soil. Alfalfa on the other hand thrives in soil that is only *slightly* acidic or *slightly* basic. Also, as previously mentioned, soil pH affects nutrient availability (Colorado State University Extension, 2011). **Nitrogen is available to plants when the pH is above 5.5. Phosphorous is available when the soil pH falls between 6.0 and 7.0.** If the acidity of the soil becomes too great however, plants cannot pick up essential nutrients from the soil. Instead, in highly acidic soil, plants are more apt to take up poisonous, toxic metals.

Soil Nutrients

The nutrients provided by soils are essential for plant growth. *Macronutrients* include **nitrogen, phosphorous,** potassium, calcium, magnesium, and sulfur, and are required in large amounts (Montana State University, 2004). *Micronutrients*, also called trace elements, include iron, manganese, boron, zinc, copper, chlorine, cobalt, molybdenum, and nickel, and are required by plants in smaller amounts (Montana State University, 2004).

Soil Conductivity

Soil conductivity is a measure of the salinity of a soil. Salt in the soil influences microbial activity. Too many salts hinder plant growth, and the more salts dissolved in solution, the higher the conductivity reading.

References:

BrainyQuote. (2014). Franklin D. Roosevelt. Retrieved from http://www.brainyquote.com/quotes/authors/f/franklin_d_roosevelt_2.html.

Colorado State University Extension. (2011). Soil pH. Retrieved from http://www.ext.colostate.edu/mg/gardennotes/222.html

Cornell University. (2014). Soil basics: biology. Retreived from http://blogs.cornell.edu/horticulture/soil-basics/soil-basics-biology/

Montana State University. (2004). Chemical analyses: solids, macronutrients. Retrieved from http://www.ecorestoration.montana.edu/mineland/guide/analytical/chemical/solids/macronutrients.htm.

National Estuarine Research Reserve System [NERRS]. (2010). Soil composition and formation. Retrieved from http://www.nerrs.noaa.gov/Doc/SiteProfile/ACEBasin/html/envicond/soil/slform.htm.

University of Michigan. (2013). Plate tectonics and soils, weathering, and nutrients. Retrieved from http://www.globalchange.umich.edu/globalchange1/current/lectures/soils/soils.html.

SOIL ANALYSIS PART 1 DATA SHEET

Name: _____

Soil Acidity – (record pH) – Follow the Directions Included with the Kit

pH of Soil Sample A on a 0 to 14 Scale   _____
pH of Soil Sample B on a 0 to 14 Scale   _____

Nitrogen (N), Phosphorous (P), and Potassium (K) Analysis   **Follow Kit**   **For each nutrient, record as Low (L), Medium (M), or High (H):**   **Directions**

Nitrogen   For Soil Sample A _____   For Soil Sample B _____ (DEMO)
Phosphorous   For Soil Sample A _____   For Soil Sample B _____ (DEMO)
Potassium   For Soil Sample A _____   For Soil Sample B _____

Soil Conductivity – Perform on Both Soils

Conductivity of Soil A in Appropriate Unit   _____
Conductivity of Soil B in Appropriate Unit   _____

Soil Analysis Part 2: Physical Properties and Indicators

Student Learning Outcomes:

1. To become familiar with basic physical properties of soil.
2. To learn about the importance of soil quality indicators.
3. To continue with a basic comparison of two different types of soil.
4. To use a USDA Soil Texture Triangle.
5. To use a USDA Soil Survey.

Discussion:

Last time, we looked at chemical properties associated with soil. Specifically, we examined pH, nutrient content, and conductivity. We also considered the fact that soils, although essentially non-renewable, are reusable if managed correctly. We pointed out that as environmentalists we need to practice good stewardship. In other words, when we use our planet's soil resources, we should leave the soil is as good a condition as we found it. Today, we will examine physical properties associated with soil including texture, bulk density, and moisture.

Soil quality is "the capacity of a specific kind of soil to function, within natural or managed ecosystem boundaries, to sustain plant and animal productivity, maintain or enhance water and air quality, and support human health and habitation" (USDA, 2001). When we examine parameters such as pH, nutrient content, conductivity, texture, bulk density, and permeability, we are examining soil quality indicators. **Farmers and scientists alike use these indicators to assess soil quality and to determine how best to manage the land. Soil quality indicators are used to determine which land-use practices are most suitable. Once soil condition is identified, the soil can then be appropriately managed in a sustainable way.**

Soil Profile

As shown in Figure 1, soil is comprised of layers: the O horizon, the A horizon, the E horizon, the B horizon, the C horizon, and the R horizon (National Estuarine Research Reserve System [NERRS], 2010). The O horizon is the top, organic layer of soil and is comprised mostly of decaying plant and animal material. It is very nutrient rich. The A horizon, or topsoil, is found just below the O horizon. **It should be noted that a high organic content promotes soil fertility and encourages plant growth. Darker soils generally have a higher level of organic matter.** The A horizon is made up of partially decayed organic matter (called humus) and mineral particles and is commonly called the topsoil. This is the layer in which seeds germinate and roots take hold. Next is the E horizon. The E horizon is the zone where leaching occurs. It therefore generally has a lighter color. The B horizon, or subsoil, is comprised of clay and mineral deposits that have accumulated through leaching. The C horizon is made of broken-up bedrock and the R horizon is comprised of unweathered parent rock (NERRS, 2010).

SOIL LAYERS

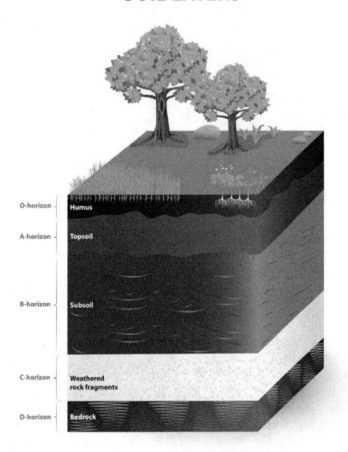

Figure 1. Soil layers.

Soil Texture

Soil texture is a measurement of the proportions of sand, silt, and clay present (NERRS, 2010). Soil texture describes the size of the soil particles. Sandy soils, although heavily aerated and easy to till, dry out quickly. The nutrients in sandy soils have a tendency to wash away or be leached below the root zone. Clay soils resist erosion and hold nutrients, but have a tendency to become waterlogged and sometimes have difficulty supporting tractors. Soils with a high silt content drain well and hold their nutrients. They are also easy to plow. Silty soils are therefore generally better for agriculture than either sandy soils or clay soils. **Loams soils, which contain an equal blend of the characteristics of sand, clay, and silt, are by far the most suitable for agriculture** (NERRS, 2010).

Bulk Density of Soil

Bulk density is the "ratio of the mass of dry solids to the bulk volume of the soil occupied by those dry solids" (Oak Ridge National Laboratory, 2010). It is the mass per unit volume for a dry soil. **If soil is very dense, roots cannot grow and nutrients and water move through the soil very slowly.** Roots in the soil tend to lower the bulk density. Aeration is poor and toxic gases have a tendency to build up in extremely dense soils.

References:

National Estuarine Research Reserve System [NERRS]. (2010). Soil composition and formation. Retrieved from http://www.nerrs.noaa.gov/Doc/SiteProfile/ACEBasin/html/envicond/soil/slform.htm.

Montana State University. (2004). Chemical analyses: solids, macronutrients. Retrieved from http://www.ecorestoration.montana.edu/mineland/guide/analytical/chemical/solids/macronutrients.htm.

Oak Ridge National Laboratory [ORNL]. (2010). Soil bulk density data. Retrieved from http://daac.ornl.gov/FIFE/Datasets/Soil_Properties/Soil_Bulk_Density_Data.html.

USDA Natural Resources Conservation Service [NRCS]. (2014). Healthy soil for life. Retrieved from http://www.nrcs.usda.gov/wps/portal/nrcs/main/soils/health/.

USDA Natural Resources Conservation Service [NRCS]. (2001). Soil quality – introduction. Retrieved from http://urbanext.illinois.edu/soil/sq_info/sq_intro.pdf.

SOIL ANALYSIS PART 2 DATA SHEET

Name: _____

Soil Profile – (describe color, texture, structure, etc. using your powers of observation)

Soil Sample A and Soil Sample B

_____

_____

_____

_____

_____

Soil Texture (sand, silt, clay, or loam?) Test Both Soils and Triangle: A and B

------------------------------	------------------------------	Soil A	Soil B
A Total Height	**Measure Directly**		
B Sand and Silt Height	**Measure Directly**		
C Sand Height	**Measure Directly**		
D Silt Height (darker layer)	**B - C**		
E Clay Height	**A - B**		
% Silt	**(D/A) X 100**		
% Clay	**(E/A) X 100**		
% Sand	**(C/A) X 100**		
Texture Triangle Results Soil A			
Texture Triangle Results Soil B			

Soil Sample - DEMONSTRATION	Soil Texture
USDA Ribbon Test Results Soil A	
USDA Ribbon Test Results Soil B	

Soil Bulk Density – (express as grams per milliliter)			
		Soil A	Soil B
Volume of Empty Can (mls)	**Bottom of Can**		
Mass of Empty Can (g)	**Bottom of Can**		
Mass of Soil & Can **AFTER** Drying (g)	**Weigh on Balance**		
Mass of Dry Soil (g)	**By Subtraction**		
Crude Bulk Density (in g/ml)	**Mass / Volume**		

Post-Lab Questions - TO BE COMPLETED ON YOUR OWN IN COMPLETE SENTENCES:
Answer in complete sentences using proper grammar.

1. Look at the following chart. If you wanted to raise alfalfa, based on your pH measurements, would the land from which Soil A was collected be suitable? Soil B? What if you wanted to raise blueberries? Shaded areas represent the suitable pH range.

Suitable Soil pH Ranges (based on table from Whitaker et al., 1959)

	4.5 – 5.0	5.0 – 5.5	5.5 - 6.0	6.0 – 6.5	6.5 – 7.0	7.0 – 7.5
Blueberries		▓	▓			
Radishes			▓	▓		
Strawberries		▓	▓	▓		
Alfalfa					▓	▓

2. Is the nutrient content of your soils low, medium, or high? Rather than using an artificial fertilizer, what other ways might an increase in soil fertility be accomplished. In other words, how might an organic farmer manage the fertility of his or her soil?

3. Soil conductivity should be kept between 40 - 400 µS (micro-Siemens). Are the soils you tested acceptable for general crop growth?

4. If you were going to plant a garden, would the soil examined today be suitable? Consider texture and bulk soil density. Use the chart that follows and justify your answer.

Relationship of Bulk Density to Root Growth – (based on data from NRCS, 2014)

Soil Texture	Ideal bulk density (g/cm^3)	Bulk densities that may affect root growth (g/cm^3)	Bulk density that restrict root growth (g/cm^3)
Sands, loamy sands	<1.60	1.69	>1.80
Sandy loams, loams	<1.40	1.63	>1.80
Sandy clay loams, loams, clay loams	<1.40	1.60	>1.75
Silts, silt loams	<1.30	1.60	>1.75
Silt loams, silty clay loams	<1.10	1.55	>1.65
Sandy clays, silty clays, some clay loams (35%-45% clay)	<1.10	1.49	>1.58
Clays (>45% clay)	<1.10	1.39	>1.47

5. Which soil overall (A or B) seems MOST suitable for agriculture? Justify your answer.

Water Quality and Fecal Coliform

> *"If there is magic on the planet, it is contained in Water."*
> - Loren Eiseley, The Immense Journey, 1957.

Student Learning Outcomes:

1. To learn about the water cycle.
2. To understand the difference between a point source and a non-point source of pollution.
3. To learn about the different categories of water pollutants.
4. To learn about fecal coliform and what its presence indicates.
5. To test a water sample for the presence of coliform and determine probable number of coliform for 100 mls of a water sample (1st stage presumptive test only).

Materials:

Water samples, lactose broth tubes, markers, 10 ml pipettes.

Discussion:

Freshwater is a renewable resource. A renewable resource is a resource that can be replenished with time. Freshwater is replenished through the natural and normal functioning of the water cycle. The water cycle is a global cycle. The main reservoir for the water cycle is the world ocean, which covers approximately three-fourths of the earth's surface.

There are four processes essential to the water cycle: evaporation, transpiration, condensation, and precipitation. First, let's consider *evaporation*. Pretend you have a beaker of water. The water contained in the beaker is made up of water molecules. Each water molecule is made up of 2 hydrogen atoms and 1 oxygen atom, hence the chemical formula H_2O. These water molecules are in motion. They bounce against the walls of the container and are reflected and redirected. They also collide with and are reflected off of one another. Some of the molecules collide against the water's surface. Sometimes the molecules collide against the water's surface with enough force to 'break free' and escape into the atmosphere (Botkin & Keller, 2010).

When water molecules escape from a body of water into the atmosphere, this is called evaporation. Eventually all of the water would evaporate from out of your beaker. If you were to place your beaker outside on a hot day, the water would evaporate even faster. Higher temperatures cause the molecules to move with additional speed which makes it easier for the molecules to break away into the atmosphere. Evaporation is driven by the sun and occurs in the oceans and in other bodies of water (and from the soil, for that matter) just as it occurs in our imaginary beaker. Evaporation is the transition from liquid water to gaseous water vapor, it occurs on planet Earth, and it is a key part of the global water cycle (Botkin & Keller, 20010).

Transpiration also contributes water to the atmosphere and is also a key part of the global water cycle. Transpiration is simply the evaporation of water from the green leaves of plants. Plants open their stomata, or small pores, during photosynthesis to allow for the transfer of gases, especially CO2 (used up during photosynthesis) and O2 (created during photosynthesis). The opening of the stomata also releases water vapor to the atmosphere. In fact, one of the primary ecological functions of forests is to return water to the atmosphere and to aid with the global water cycle (Botkin & Keller, 2010).

Water moves to the atmosphere through evaporation and transpiration. But how does it return to earth? Water returns to earth through condensation and precipitation. Condensation is the opposite of evaporation. When air cools, the water molecules in the air start to slow down. When they slow down, they move from the gaseous water vapor state back to the liquid state. ***Condensation*** is the third key component of the water cycle. When gaseous water condenses in the atmosphere, clouds form. Clouds in the atmosphere are made up of tiny droplets of water. When the cloud particles become too heavy to remain suspended in the atmosphere, they fall to earth. This of course is ***precipitation***. We usually think of precipitation as rain, but remember that precipitation can also descend from the atmosphere as sleet, ice, snow, and hail. Precipitation is the fourth key process within the water cycle. Once water has returned to earth, it runs across the land back into rivers and lakes and large bodies of water. Some water seeps into the earth and forms groundwater reservoirs. All of this water ultimately flows back to the sea (Botkin & Keller, 2010).

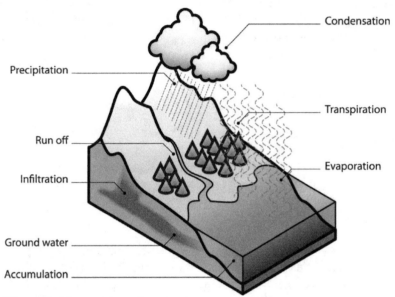

Figure 1. The water cycle.

It is fortunate for us that water is a renewable resource, because in reality, even though the earth is covered in water, freshwater, the only water that is suitable for drinking, accounts for only about 3% of that. Most of this 3% is tied up in the icecaps and in glaciers, which leaves only a very small percentage coming from lakes, rivers, streams, wetlands, and groundwater (Raven & Berg, 2001).

Environmental scientists often point out that freshwater is renewable only to a point. **Our supply of freshwater is threatened by both wasteful practices and various human activities.** For example, when human beings cut down large sections of tropical rainforest, they are destroying the ability of those forests to release water vapor into the atmosphere through transpiration. This can alter and disrupt local, regional, and global weather patterns. Another human activity that disrupts the water cycle is the pumping of groundwater out of the earth in order to irrigate crops. This causes the soil to dry out much faster than it would normally and it decreases groundwater supplies. This is a problem because underground aquifers generally refill very slowly.

Pollution also threatens our global supply of freshwater. For example, when our automobiles and power plants release certain gases into the atmosphere, these gases can combine with precipitation to form sulfuric acid and nitric acid. Acidic solutions literally rain down on the earth, entering bodies of water and harming plant and animal life. When vegetation is destroyed by acid rain, this again reduces the amount of water being returned to the atmosphere through transpiration.

Acids and other caustics are not the only ***inorganic chemicals*** that find their way into our freshwater supply. Toxic metals are often produced as by-products during industrial processes. The metals produced during certain types of mining and ore processing are often radioactive and can enter bodies of water through surface run-off. ***Organic chemicals*** such as pesticides, detergents, and oil are also serious water pollutants. Organic chemicals are often used in industry to manufacture pharmaceuticals and plastics. Organics enter bodies of water through agricultural run-off, through industrial discharges, and through the improper disposal of household wastes (Cunningham & Cunningham, 2009).

Nitrates and phosphates also contaminate our freshwater. Nitrates and phosphates are found in agricultural waste such as fertilizers and manure. Nitrates and phosphates are ***plant nutrients***. They stimulate the rapid growth of plants and algae. This can result in a condition called eutrophication. When the plants and algae die and decompose, they cause the levels of dissolved oxygen in the water to drop, thereby harming biodiversity and degrading water quality. Eutrophication will be discussed in greater detail elsewhere in this lab book. It should be pointed out that any material that stimulates oxygen consumption by decomposers is considered an ***oxygen-demanding waste***. Sewage for example is an oxygen-demanding waste (Cunningham & Cunningham, 2009).

Sediments such as dirt, sand, and silt can also degrade water quality. Sediments enter bodies of water through erosion. Erosion can be caused naturally by wind or by water. Erosion and the disturbance of the soil are also caused by development projects, by agriculture, by road construction, and by logging and mining. Once sediments enter a body of water they can create murky conditions and raise water temperatures, thereby disrupting fish reproduction and harming biodiversity. Sediments can also clog fish gills, smother fish eggs, and blanket fish nests when they settle to the bottom (Cunningham & Cunningham, 2009).

It should be noted that **heat** is even classified as a water pollutant under the federal government's Clean Water Act. Power plants and industries sometimes take in water from rivers and streams to cool their equipment. When they release this water back into the river or stream, the water has higher heat content, and this results in thermal pollution which can harm aquatic organisms (Cunningham & Cunningham, 2009).

Whatever the classification, water pollution can be either point source pollution or non-point source pollution. ***Point sources*** of pollution come from specific sources, are well-defined, and are easily identified. A drainage pipe that empties into a stream is a good example of a point source. A ***non-point source*** on the other hand covers a larger area, is more diffused, and is not as easily identified or measured. The storm water run-off from a large parking lot is an example of a non-point source pollutant (Cunningham & Cunningham, 2009).

The pollutant that will be focused on in this week's laboratory exercise is classified as an ***infectious agent.*** Infectious agents are disease-causing pollutants. These pollutants contain pathogens. Pathogens are disease-causing organisms such as parasites, viruses, and bacteria. Infectious agents come from human fecal matter, from farm animal manure, from sewage, and from leaky septic systems (Cunningham & Cunningham, 2009).

Coliform bacteria are a family of bacteria found naturally in soils, water, plants, and animals. This family of bacteria is mostly harmless. Some coliform bacteria are even beneficial. For example, colifom bacteria aid animals in the digestion of food. One type of coliform bacteria is fecal coliform bacteria. Fecal coliform bacteria are present in the intestinal tracts of warm-blooded animals. Fecal coliform bacteria in swimming pools and holes may or may not make a person sick; fecal coliform in drinking water is of greater concern (ingestion of *Escherichia coli* for example causes abdominal cramps and diarrhea). The presence of fecal coliform in a body of water indicates that human and/or animal waste and raw sewage has been introduced to the body of water. Such waste is known to spread infectious agents. In other words, **the presence of fecal coliform is a clue that there could be other, even more serious infectious contaminants present in the water.** For this reason, fecal coliform is considered an environmental indicator (Cunningham & Cunningham, 2009).

Today, we will test water for the presence of coliform (or of some other lactose fermenting bacteria) using Lactose broth tubes with Durham tubes. Today's test constitutes a ***presumptive test***. A positive presumptive test shows that coliform bacteria are more than likely present. We should therefore assume for any water sample giving a positive presumptive response that the water tested is not fit for drinking. As for whether or not *fecal* coliform bacteria are present, more testing would be required before we could say with any certainty.

References:

Botkin, D.B. and Keller, E.A. (2010). *Environmental science – Earth as a living planet*. Hoboken, NJ: John Wiley and Sons

Cunningham, M.A. and Cunningham, W.P. (2009). *Environmental Science: A Global Concern*. Wm. C. Brown Publishers: USA.

Adapted from a standard technique (the multiple-tube fermentation technique) outlined in a variety of sources, including, but not limited to:

Cappucino, J., and Sherman, N. (1998). *Microbiology: A laboratory manual*. Benjamin-Cummings Pub Co: USA.

Rockett, C., and Van Dellen, K. (1993). *Laboratory manual for living in the environment, environmental science, and sustaining the earth*. Belmont, CA: Wadsworth Publishing Company.

Enger, E. and Smith, B. (2000). *Field and laboratory exercises in environmental science*. McGraw Hill: USA.

Water Quality and Dissolved Oxygen

Student Learning Outcomes:

1. To learn about dissolved oxygen and percent oxygen saturation.
2. To learn about the relationship between dissolved oxygen and temperature.
3. To become familiar with the concept of biochemical oxygen demand.

Materials:

Percent saturation chart, straight edges

Discussion:

Dissolved oxygen is the amount of oxygen gas in solution in a body of water (Environmental Protection Agency [EPA], 2012). **If a body of water is healthy, the dissolved oxygen content of that body of water will likely be at acceptable levels.** If a body of water has very low dissolved oxygen content, that body of water is probably not healthy. For example, high levels at 8-10 milligrams per liter (mg/L) are considered good, while levels below 2 mg/L are low and troublesome to aquatic organisms (EPA, 2012). Just as land organisms (ourselves included) need oxygen to live and carry out normal life processes, aquatic organisms also need oxygen to live and function normally. Fish especially need oxygen. Land organisms respire with their lungs (respiration requires oxygen), and fish respire with their gills.

Where does oxygen in the water come from? Oxygen enters a body of water in a natural exchange process with the atmosphere, especially when the water is stirred up or disturbed. Oxygen in the water also comes in large part from aquatic plants engaging in photosynthesis (EPA, 2012). Consider the following equation for the process of photosynthesis:

$$\text{Sunlight} + 6H_2O + 6CO_2 \text{ ----------> } C_6H_{12}O_6 + 6O_2$$

During photosynthesis, plants combine water with carbon dioxide in the presence of sunlight to produce a sugar called glucose ($C_6H_{12}O_6$). The byproduct of photosynthesis is oxygen, which is of course necessary for life. Oxygen is therefore introduced into the water. Consider also the following equation for the process of respiration:

$$C_6H_{12}O_6 + 6O_2 \text{ ----------> } 6H_2O + 6CO_2 + \text{energy}$$

Respiration is the opposite of photosynthesis. While photosynthesis is the production of glucose, respiration is the breaking down of glucose by organisms to release energy for life processes. Respiration requires oxygen.

Human beings can disrupt and alter the normal composition of any body of water by polluting that body of water. For example, if excess nutrients are added to a body of water from sewage or from fertilizer run-off, these nutrients can rapidly stimulate plant and algae growth. The explosive growth of algae from nutrient enrichment is called an algal bloom. Algal blooms are carpets of algae that form on the surface of the water. Algal blooms on the surface of the water block out the sunlight necessary for photosynthesis. When algal blooms block out sunlight, the photosynthetic process is disrupted in aquatic plants and less oxygen (the by-product of photosynthesis) is produced (EPA, 2012).

Plants and algae engage in photosynthesis during the day; they also engage in respiration, even during the night when photosynthesis ceases (EPA, 2012). Less oxygen being produced by photosynthesis during the day and excessive amounts of plants and algae respiring and using up oxygen serves to decrease the overall levels of oxygen in the body of water. Eventually, the plants and the algae produced by the nutrient-rich conditions die. The dead plants and algae sink to the bottom and are decomposed by bacteria. Bacterial populations increase as a result of the excessive amounts of dead organic material now available. Bacteria consume oxygen when they break down organic matter in order to obtain energy. High rates of decomposition can result in an overall decrease in dissolved oxygen. Additional nutrients equals prolific plant and algae growth, which results in less dissolved oxygen due to a decreased rate of photosynthesis, increased respiration, and an increase in decomposition. Fish, invertebrates, and aquatic vegetation need dissolved oxygen in the water to survive and without sufficient levels, fish kills and massive die-offs of other aquatic life can result (EPA, 2012).

A natural increase in the amount of plant nutrients (such as nitrogen and phosphorous) in a body of water is called ***eutrophication***. When human activities bring about a drastic and rapid increase in the amount of nitrogen and phosphorous in a body of water, this is called cultural eutrophication. In short, cultural eutrophication produces rapid plant and algal growth, which can have adverse effects on a body of water and can result in the degradation of water quality.

Any organic waste added to a body of water can lead to a decrease in dissolved oxygen. Decomposers such as bacteria break down organic wastes and convert the wastes to energy. In the process, these decomposers consume oxygen. The amount of oxygen they consume is called the ***biochemical oxygen demand***, or BOD (EPA, 2012). The more oxygen consumed by the decomposers (the higher the BOD), the less oxygen available to fish and other forms of aquatic life. Any pollutant that leads to a reduction in the amount of dissolved oxygen is called an oxygen-demanding waste. It should be noted that the temperature of the water greatly influences the amount of dissolved oxygen that body of water can hold as well. In general, warmer water holds less dissolved oxygen (EPA, 2012).

References:

Environmental Protection Agency [EPA]. (2012). Dissolved oxygen and biochemical oxygen demand. Retrieved from http://water.epa.gov/type/rsl/monitoring/vms52.cfm.

Famous Quotes About. (2014). H. Ross Perot. Retrieved from http://www.famousquotesabout.com/quote/The-activist-is-not/505145.

Based in part on a classic experiment traditionally performed at UT Chattanooga and on an exercise prepared for the Environmental literacy Council (as outlined in)

Wilkins W., Hierholzer J.C., Bock C., Laderer P., and others. (n.d.). *Environmental Studies – Laboratory Manual for EST 150 Students.* (Unpublished).

Palmer L. (2008). *Dissolved Oxygen.* The Environmental Literacy Council. Retrieved from http://www.enviroliteracy.org/article.php/1159.html

Dissolved Oxygen Determination with the Winkler Method

"Water, thou hast no taste, no color, no odor; canst not be defined, art relished while ever mysterious. Not necessary to life, but rather life itself, thou fillest us with a gratification that exceeds the delight of the senses."

-Antoine de Saint-Exupery (1900-1944), *Wind, Sand and Stars*, 1939

Student Learning Outcomes:

1. To learn about titration as an analytical technique.
2. To perform titrations for water quality analysis.
3. To collect water samples in a BOD bottle.
4. To conduct the classic Winkler method for the determination of dissolved oxygen.

Discussion:

You have already learned about the importance of dissolved oxygen in a body of water and its usefulness as a water quality parameter. Today, you will learn a classic technique for determining the concentration of dissolved oxygen in creeks and streams and get hands-on experience performing a titration for the purpose of water quality analysis.

Titration is a powerful and useful analytical technique. It is a technique that you are likely to encounter again if you spend any considerable amount of time in a laboratory. The concept of the titration is relatively simple. A reagent of known concentration is added to a sample of unknown concentration until an indicator changes color. The color change marks the 'endpoint' of the titration. At the endpoint, the amount of reagent added is equivalent to the amount of the compound of interest in the sample. In other words, the amount of reagent added is useful in determining or calculating the concentration of the sample. Today, you will collect water samples in a ***BOD bottle***. You will then add a series of reagents to 'fix' the amount of oxygen in the sample and a small amount of sulfuric acid to dissolve any precipitate. You will then titrate the sample with standardized sodium thiosulfate solution until a color change from bright gold to pale yellow occurs. A few drops of starch solution will turn the sample dark blue. You will then add the sodium thiosulfate drop wise until a single drop (or better yet, a half-drop) turns the solution clear. This is the endpoint of the titration. One milliliter of sodium thiosulfate is equivalent to 1 milligram per liter (or ppm) of dissolved oxygen in the sample (Laval, 2006; Bruckner, 2013).

Example: An environmental science student scrubs and cleans her burette with a dilute soap solution. She rinses the burette thoroughly (at least 3 times with tap water and 3 more times with distilled water before finally rinsing with a small amount of sodium thiosulfate). After discarding the rinse solution, she fills the burette with the sodium thiosulfate. She briefly opens the stopcock and releases a small amount of the titrant into a waste beaker in order to blow out any bubbles in the tip of the burette. She then reads the volume of the sodium thiosulfate solution by noting the bottom of the meniscus. This is the initial volume. She records this volume with the appropriate unit out to two decimal places (milliliters). Her initial volume is 0.38 milliliters.

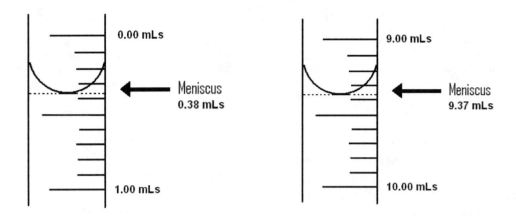

She opens the stopcock and adds titrant to the sample very slowly and with constant swirling. She is very careful to make sure that none of the sample splashes up on the side of the Erlenmeyer flask. If any does splash up, she simply   washes down the sides of the flask with a small amount of distilled water. As she approaches the endpoint, she adds the titrant drop by drop. Ideally, a single drop should produce a permanent color change (in this case, a single drop at the endpoint should make the blue solution turn colorless). At the endpoint, the student reads the buret yet again and determines the final volume to be 9.37 milliliters. To determine the total volume delivered (the total volume needed to achieve a complete reaction), she subtracts the initial volume from the final volume:

9.37 mls - 0.38 mls = 8.99 mls

Since 8.99 mls of reagent were required to reach the endpoint, the student knows that the dissolved oxygen content of her water sample is 8.99 mg/L or 8.99 ppm.

Procedure:

1. Immerse the BOD bottle directly into the stream. Make sure the bottle is completely filled (with no air space at the top) and cap the bottle. Be careful not to agitate the water since any agitation could alter the oxygen levels.

2. Fix the oxygen levels immediately by adding the appropriate fixing reagents. Add one milliliter (below the surface with a small pipette) $MnSO_4$ solution and one milliliter alkali- iodide solution. Mix gently by inversion and allow the formation of a brown precipitate. Let the precipitate settle. Collect and repeat with another sample.

3. Place the bottles in a cooler and return to the lab.

4. Sometime within the next 48 hours, continue with the experiment. First, add one milliliter of concentrated sulfuric acid (H_2SO_4). The precipitate will dissolve and   the solution will become gold in color.

5. Add 2/3 of the gold solution (201 milliliters) to an Erlenmeyer flask.

6. Fill the buret with sodium thiosulfate. Open the stopcock and release a small amount into a waste beaker to remove any air bubbles and note and record the initial volume of sodium thiosulfate.

7. Add the titrant slowly and with stirring until a faint yellow color is achieved. Record the new volume on the data sheet.

8. Add three drops of starch solution to the Erlenmeyer flask and mix gently to bring about a dark blue color.

9. Add more sodium thiosulfate VERY SLOWLY drop by drop. At the endpoint, the solution will be colorless (a single drop or half of a drop should be sufficient to reach the endpoint).

10. Record the final volume of sodium thiosulfate on the data sheet.

11. Determine the total volume delivered. The total volume delivered in milliliters is equivalent to the concentration of dissolved oxygen in milligrams per liter (or ppm).

12. Repeat with the second sample. The second trial should go much faster and run much more smoothly since you already know the approximate location of the endpoint.

References:

American Public Health Association. (2006). *Standard Methods for the Examination of Water and Wastewater*. Retrieved from http://standardmethods.org/.

Bruckner, M. (2013). *The Winkler Method - Measuring dissolved oxygen*. Microbial Life Educational Resources. Available at http://serc.carleton.edu/microbelife/research_methods/environ_sampling/oxygen.html

Water for the Ages. (2014). De Saint-Exupery. Retrieved from http://waterfortheages.org/2007/10/29/water-words-quote-of-the-day/

Laval, B. (2006). *Determination of dissolved oxygen (Winkler Method)*. Department of Civil Engineering, University of British Columbia.

University of Toronto at Mississauga. (n.d.). *Laboratory in physiology lab manual - Oxygen consumption: effects of body size and temperature in aquatic animals*. Available http://www.erin.utoronto.ca/~w3lange/bio309/labs/Bio30902Cons…

DISSOLVED OXYGEN DETERMINATION WITH THE WINKLER METHOD

NAME: _____

	Trial 1 - Sample 1	Trial 2 - Sample 2
Initial volume (sodium thiosulfate)		
Volume at pale yellow (sodium thiosulfate)		
Final volume (sodium thiosulfate)		
Total volume (sodium thiosulfate)		
Dissolved oxygen in mg/L		
Average of dissolved oxygen for the two samples		

Post-Lab Questions:
Answer in complete sentences using proper grammar.

1. Make some statement about the overall quality of the stream based on your determination of DO content .Would you expect the stream tested to be conducive to aquatic life?

2. At the time of sampling, your instructor checked the stream with a DO meter. Does your determination agree with the instructor's readings? If not, why not? Which do you trust?

3. Do some research and try to explain in simple terms exactly what is happening chemically during the titration. Those of you that have had chemistry may have an easier time with this question.

Storm Water Runoff

"Water is the one substance from which the earth can conceal nothing; it sucks out its innermost secrets and brings them to our very lips."

-Jean Giraudoux (1882-1944), The Madwomen of Chaillot, 1946

Student Learning Outcomes:

1. To map and calculate the area of a parking lot.
2. To calculate the volume of runoff from the parking lot.
3. To map the route of surface runoff to the nearest body of water.

Discussion:

Stormwater runoff occurs when rainwater flows over the ground and impervious surfaces into a storm sewer system or into a body of water such as a river, lake, or stream (Environmental Protection Agency [EPA], 2012). As rainwater flows over the landscape, it picks up chemicals, sediment, and debris which can be harmful to water supplies if not treated (EPA, 2012). For example, agricultural runoff can contain pesticides, fertilizers, and sediment. Urban storm water runoff may contain debris and chemicals such as gasoline, oil, road salt, and heavy metals.

Runoff from large areas of pavement is particularly likely to contain pollutants, since none of the water or pollutants can be absorbed through the pavement. In addition, urbanization adversely affects water quality by increasing the volume of surface runoff while decreasing runoff times. When it rains, more water runs off at a higher speed because it is not absorbed into the ground. Hence, potential pollutants are transported more quickly from the land to the receiving water, which is called **shock loading**.

In Chattanooga, the sewer system is combined with storm drains. The wastewater/stormwater mix travels through pipes in the city to the Moccasin Bend Wastewater Treatment Plant where it is treated. Because of this, it is especially important for businesses and residents to take measures to reduce the amount of stormwater that goes into the antiquated and strained wastewater treatment system (Smith, 2014). Ways to reduce stormwater runoff include:
- Permeable pavement
- Rain barrels
- Green streets that capture and store rainwater
- Green roofs

View the YouTube video "Stormwater Runoff 101": http://www.youtube.com/watch?v=eozVMJCYHCM and take notes below:

 NOTES

Procedure:

1. Work in teams of 3-4 individuals.

2. Examine the map of the parking lot. Decide what measurements will be taken in the field to calculate the total area of the parking lot.

3. Taking measurements

   - Take measurements of the parking lot using 50-meter tapes. Make the measurements along the edges of the parking lot. Record the measurements (in meters) on the map with a dry-erase pen.

   - Estimate the direction of stormwater runoff. Draw arrows on the map to show the direction of flow.

   - Record any evidence of potential non-point source pollutants, and indicate on the map.

4. Calculate the area of the parking lot

   - Diagram the parking lot on graph paper (to scale).

   - Determine the area that one square on the graph paper is equal to. For example, if one side of the parking lot is 30 meters, the map should be drawn so that each square is equivalent to 1 meter. Indicate this relationship in a map legend.

   - To find area, count the number of squares contained within the perimeter of the map, and multiply this by the area one square represents.

   Area = _____503_____ m^2

C. Calculate the average amount of rain that falls on the parking lot in one year.

   Follow instructions on the next page.

References:

Environmental Protection Agency [EPA]. (2012). National Pollutant Discharge Elimination System Stormwater program. Retrieved from http://cfpub.epa.gov/npdes/home.cfm?program_id=6.

Smith, E. (2014, June 15). Flying Squirrel builds unique road in Chattanooga to deal with stormwater runoff. *Times Free Press*. Retrieved from http://www.timesfreepress.com/news/2014/jun/15/flying-squirrel-builds-unique-road-chattanoo/.

Tennessee Valley Authority [TVA]. (1993). Environmental Resource Guide.

Geographic Information Systems

> *"I am told there are people who do not care for maps, and I find it hard to believe."*
> - Robert Louis Stevenson, Treasure Island

Student Learning Outcomes:

1. To perform simple GIS mapping.
2. To learn about the contamination of Chattanooga Creek.

Materials:

Access to computers with high-speed internet connections

Discussion:

Geographic Information Systems (GIS) is a computer system that allows users to manage data and link that data to geographic locations. GIS allows the user to organize and view data spatially on a map (Scurry, n.d). While it is often helpful to locate an individual feature on a map (today you may wish to find your house or your college or university), it is often more helpful to look at a distribution of features and attempt to establish relationships between those features and other features.

With GIS, data is analyzed from a geographical perspective, and information is collected in layers so that it is easier to make generalizations and draw conclusions about that information. For example, an environmental scientist might map out areas with heavy concentrations of industry. It then might be possible to map out areas of heavy environmental degradation in order to see if there is any correlation between industrial areas and areas of heavy pollution. An environmental scientist or environmental lawyer might then be able to map out low-income neighborhoods to determine whether or not low-income families are more likely to be situated in locations near industry where they have to bear a disproportionate amount of an environmental burden. Health information could then be added to the map and checked to possibly establish a correlation between areas of environmental degradation and high infant mortality rates. All of this information (industry concentration, environmental degradation, the location of low-income residences, and infant mortality rates) could be viewed and managed in layers on a map, thereby making analysis of the information easier. When data is displayed spatially on a map, patterns begin to emerge. These patterns are sometimes easier to see in map form as opposed to finding patterns in page after page of charts, graphs, and tables.

Perhaps a developer wants to build a new set of condominiums in a desirable part of the city. He or she could use GIS to identify how close the proposed condominiums would be to environmental concerns that might hinder construction. The developer could analyze the development site on a map alongside the location of protected wetlands, for example. Perhaps a county government could map out areas of methamphetamine production in the county and then increase the presence of law enforcement in those areas. Perhaps a global health organization could use GIS to map the spread of tuberculosis alongside population density to determine where the disease is most likely to strike next. The possibilities for GIS as a tool in the field of environmental science (and obviously in other disciplines) are limitless. Employers of all kinds are especially drawn to graduates of colleges and universities with marketable GIS skills. Hopefully, today's introduction will spark your interest in GIS. Hopefully, you will go on to acquire formal GIS training, thereby making yourself more appealing to potential employers.

Today, you will familiarize yourself with GIS mapping by utilizing the Hamilton County GIS Department's Online Mapping Site located at www.hamiltontn.gov.

Procedure:

<u>A. Contour Line Map</u>

1. Go to www.hamiltontn.gov. Click on <u>Maps/GIS Data</u> and then select <u>Map Online</u>. Then click on the Hamilton County Mapping box. You will see a map of Tennessee's Hamilton County. Click Agree.

2. If you live in Hamilton County, locate your place of residence. Go to Find an Address in the toolbar and type in your street address and hit Locate. If you do not live in Hamilton County or do not want to locate your place of residence, use a Hamilton County Phone Book and simply pick an address at random.

3. To zoom, select Zoom To. Now select Identify, select the first identify tool, and click on the selected property. You will see all kinds of information about the owner, the sale date, the longitude and latitude, etc. Now close all of the pop-ups.

4. You can then select Topo in the tool bar. Note any roadways, driveways, and parking lots. Remember that run-off from parking lots is a significant source of non-point source water pollution. The contour lines will be in brown. Each brown contour line somewhere along its length will have a contour number representing the 3-dimensional altitude of the earth's surface in feet. Consider the following as an example:

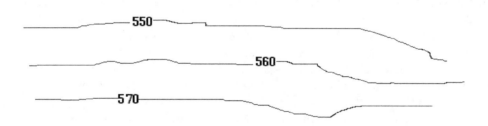

      These are contour lines. These contour lines run mostly parallel to each other. The line on the very bottom represents 570 feet. The top line represents 550 feet. Therefore, if these contour lines with their contour labels represent 3-dimensional elevations on a map, then the area of land represented by the bottom line is around 20 feet higher than the area of land represented by the top line in 3-D space. Now consider the following map created using the old Hamilton County GIS Map Maker:

Constructed with the Hamilton County GIS Mapmaker, 2007

Find the parking lot at the bottom of the map. The black arrow is pointing at the parking lot. Note that the contour line on the far right of the map (nearest the arrow) indicates 720 feet. The next line over runs through the parking lot and indicates 710 feet. The next line over indicates 700 feet and the next line over (the one closest to Shepard Road) indicates 690 feet. From right to left on this map, the elevation changes by around 30 feet. Now again, consider the parking lot of interest. During a heavy rain, which way will storm water run-off flow in the parking lot? You should be beginning to see how GIS could potentially be useful for environmental applications.

5. Print your map by selecting Print (Landscape Mode). Name the map *Contour Line Map*.

B. Arial Photo of Cemetery

1. Close any open pop-ups. Locate 615 McCallie Avenue and then close the Find an Address window.

3. Left click and hold down to use the "hand" icon to drag an item of interest to the center of the screen. Try to position your view over East 5th Street. Try to locate the Holt Science Building which sits on 5th Street.

4. Across from Holt, you should see a large open space representing three cemeteries (the general cemetery, the Confederate Army Cemetery, and the Mizpah Congregation Cemetery) that together make up the larger cemetery. Position your view so that you can see the cemetery.

5. Now select Aerial. This feature lets you see an actual photographic image of the area. Practice zooming in and out. You should be able to see the actual trees, the headstones, and the monuments in the cemetery.

6. Print your map by selecting Print (Landscape Mode). Name the map *Cemetery*.

C. Floodplain of Chattanooga Creek

Case Study:

The history of the South Chattanooga area and Chattanooga Creek is complex. Foundries, coke furnaces, chemical companies, wood preserving plants, and tanning and textile plants have thrived and prospered there for over 100 years. And over the years, many of these plants and industries have used the Chattanooga Creek area as a hazardous substance dumping ground. The most notable violator was the Tennessee Products Corporation (TPC), one of several coal carbonization facilities that operated there from 1918 through the 1970's. The TPC produced massive amounts of coal tar (starting in the 1920's) for a period of nearly forty years. The U.S. Government bought the TPC in the 1940's. In its support of the war effort, the TPC, under the direction of the U.S. Government, doubled the production of coal products and coal tar waste. Most of the coal tar (formed in the industrial transformation of coal into coke) was dumped into and along the banks of Chattanooga Creek. This likely contributed to the production of physically treacherous coal tar deposits, located primarily along a one-mile stretch of creek between Hamill Road and 38th Street. The northeast and northwest tributaries flowing from the coke plant into the Creek were contaminated from at least the early 1970's through 1987 (Reynolds, 2003).

The coke plant facility likewise operated and maintained a sewer line (used by both the Chattanooga Coke and Gas Company and the Tennessee Products Corporation) that emptied directly into Chattanooga Creek from at least the year 1944. The coal tar deposits and the chemical and industrial discharges, along with solid waste, litter, bacteria, and sewage made Chattanooga Creek one of the most polluted streams in the South and a serious potential health hazard. The industrial contaminants present in the Chattanooga Creek area (toxic metals and organic compounds) were possibly transferred from the Creek itself to the various neighboring parks, playgrounds, and gardens through wind, rain, **flooding**, and other normal weather phenomena. Industrial runoff has also possibly played a part in the contamination of the South Chattanooga Area. Much of the groundwater in the area is contaminated and the industrial contaminants were in all probability also transferred to the parks, playgrounds, and gardens through the normal flow of groundwater (Reynolds, 2003).

How did the Chattanooga Creek residents come to live in such a contaminated area? Inexpensive property was set aside in the South Chattanooga Creek area for the construction of several public housing projects for low-income families. Many Chattanooga Creek residents, mostly African-American, were relocated to this inexpensive highly contaminated area when their homes were seized by the government to make room for a major highway construction project to revitalize downtown Chattanooga. At one time, it was reported that a few homeless people also dwelled close to the Creek in order to utilize it for bathing, washing, and drinking. Chattanooga is attempting to deal with these environmental inequities. As a partial result, President Clinton's Council on Sustainable Development named the city of Chattanooga one of the Outstanding U.S. Communities in 1993. However, the city of Chattanooga is far from eradicating the environmental inequities that plague the Chattanooga Creek community. The elderly, the homeless, and the poverty-stricken usually lack the financial and political power necessary to halt the environmental inequities that are often thrust upon them. This is especially true in the Chattanooga Creek community, with the area's environmental burden falling squarely upon these three groups of people. Clearly, the basic principles of the environmental justice movement have been violated and are still being violated in the South Chattanooga community (Reynolds, 2003).

In 1983, the state of Tennessee declared Chattanooga Creek unfit and unsafe for human use. At approximately the same time, community petitions prompted the Tennessee Valley Authority to study the ambient air quality in the area. Then, in 1991, EPA, along with the state of Tennessee and the state of Georgia announced plans to address the extensive pollution and contamination of the Chattanooga Creek area. The initial part of this plan was to put a stop to ongoing contamination of the Creek. Ongoing contamination was still a serious problem as of 1991. Following stoppage of the ongoing pollution and illegal dumping, EPA's intent was to move into the cleanup stage (Reynolds, 2003).

During this time, the Chattanooga Creek community members continued to voice their concerns and get organized. Most notably, Mr. Milton Jackson along with other community members formed S.T.O.P. (Stop TOxic Pollution) in 1992 in Alton Park and through the utilization of the media helped attract the assistance and involvement of Green Peace, the Agency for Toxic Substances and Disease Registry (ATSDR), and Senator Al Gore Jr. The community petitioners from 1983 and Mr. Jackson's S.T.O.P. group were at this point striving for environmental justice. Perhaps heeding the call of local environmental justice activists, EPA fenced off part of the Creek to restrict public access in 1993. Finally, in 1995, based on an EPA study of Chattanooga Creek and a Public Health Advisory issued by ATSDR on the dangers of the coal tar deposits, the Tennessee Products Site (so named after the Tennessee Products Corporation) was placed on the federal government's National Priority List as a Superfund remediation site. EPA decided that a non-time critical removal action (one in which action may be delayed more than six months) should be undertaken to mitigate the immediate health hazard posed by the coal tar material present at the Tennessee Products Superfund Site (Reynolds, 2003).

Phase I of the Superfund cleanup began in June of 1997. It was indeed a non-time critical removal action (the cleanup began two years after declaration) focusing on the removal of the bulk coal tar deposits located primarily in the Creek, near the Chattanooga Creek floodplain east of the Oakview Residential area and north of Hamill road, and also just northeast of the former location of the Tennessee Products Corporation. Phase I was completed in October/November of 1998 and resulted in the removal of 25,300 cubic yards of coal tar. Also 1,150 cubic yards of pesticide-contaminated sediment were removed. Following the Phase I cleanup, the state of Tennessee entered a remedial investigation phase and collected composite samples from the Chattanooga Creek floodplain and investigated the presence of polycyclic aromatic hydrocarbons in and around the Creek. The Phase II cleanup focused on the smaller tar beds still present as well as on additional contaminated sediments located in the Creek and along the floodplain of the Creek (Reynolds, 2003).

References:

Reynolds, B.R. (2003). Cumulative Community Contamination and the Comprehensive Environmental Response, Compensation, and Liability Act (Unpublished master's thesis). University of Tennessee at Chattanooga, Chattanooga, Tennessee.

Scurry, J. (n.d.). What is GIS?. Retrieved from http://www.nerrs.noaa.gov/doc/siteprofile/acebasin/html/gis_data/gisint2.htm

The British Cartographic Society. (2014). R.L. Stevenson, R. L. Retrieved from http://www.cartography.org.uk/default.asp?contentID=967.

Instructions:

1. Close all open windows and pop ups. Find a new address. Locate 2500 Market Street. Again, close any open windows.

2. You should now see Howard Elementary and/or Howard School of Academics and Technology. Zoom in or out if necessary for a good view. Note the football field with the running track around it. The body of water below Howard School is Chattanooga Creek. Find Chattanooga Creek (keeping the Howard School in view).

3. Select Layers and expand Layers. Now expand FEMA. Under FEMA, check Floodway and then check FEMA. The flood plain of the creek will be highlighted in light blue. The floodplain is the land area of a river, stream, or creek that floods when the banks are exceeded.

4. Now under FEMA check 100 Year Flood. According to the Federal Emergency Management Agency, the 100 Year Flood is defined as a flood event of a magnitude expected to be equaled or exceeded once on the average during any 100-year period.

5. Now check 500 Year Flood. The 500 Year Flood is defined as a flood event of a magnitude expected to be equaled or exceeded once on the average during any 500-year period. Zoom out so that you can see the extent of the potential flood area. Close any open windows.

You will see that much of the floodplain of the highly contaminated Chattanooga Creek comes into contact with the Howard School. Contaminants most likely spread by the flooding of the Creek have been found at the Gaston Stadium Field at Howard High School and the Western Football Practice Field, which is also at Howard. The Piney Woods Elementary School Playground and the Edinburg Garden located at 4902 Edinburg Street are also contaminated. Data were collected at these sites for a variety of metals, including arsenic, lead, and chromium, and also for several polycyclic aromatic hydrocarbons (PAHs) and polychlorinated biphenyls (PCBs). These data indicate that in addition to the toxic pollutants contained in the Creek, there are also toxic pollutants in the floodplain and within the Chattanooga Creek community itself. The toxic pollutants present in the playgrounds, parks, and gardens have in part been spread by flooding and may be endangering the residents living in the Chattanooga Creek watershed.

6. Print your map by selecting Print (Landscape Mode). Name the map *Howard High*.

7. Write your name on all three maps and turn them in as your "data sheets" for this activity.

Solid Waste

> *"I had always suspected that one could build an entire house from what went into the landfill, and, sure enough, it's true."*
>
> - Dan Phillips, American Builder and Designer

Student Learning Outcomes:

1. To know about municipal solid waste.
2. To analyze sample MSW.
3. To conduct a landfill assessment.
4. To assess personal effects on the solid waste stream.

Materials:

Scales, bags of sample municipal solid waste

Discussion:

Solid waste is any unwanted or discarded material also known as garbage, trash, or refuse. In the United States, billions of tons of solid waste are produced each year and most of this waste is generated by mining, industrial, and agricultural activities. **Municipal solid waste** (MSW) is material discarded by homes and businesses (Figure 1, EPA, 2015). This constitutes a small fraction of the total solid waste produced; however, it is actually a large amount of waste to deal with each year –about 254 million tons. In fact, the average person in the U.S. produces about 4.4 pounds of waste each day (EPA, 2015). Most MSW is biodegradable or recyclable material and about 35% of MSW is recycled or composted in the United States (EPA, 2015). The good news is that over the last few decades, the general of waste per person per day has gone down to its lowest rate since the 1980's. Also, the recycling rate is at its highest rate ever (EPA, 2015). Waste that is not recycled or composted is managed by burying in a landfill or burning at an incinerator.

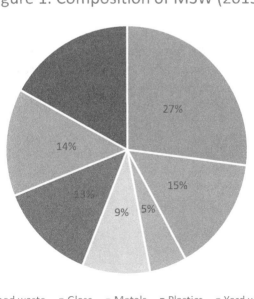

Figure 1. Composition of MSW (2015)

■ Paper ■ Food waste ■ Glass ■ Metals ■ Plastics ■ Yard waste ■ Other

Landfills

Landfills are basically places to bury trash; but they are carefully engineered. About 54% of MSW is buried in landfills (EPA, 2015). Solid waste is spread out in thin layers, compacted by bulldozer, and covered daily with soil or plastic. After a landfill has been used to capacity, it is capped with clay and plastic and can be landscaped.

There are drawbacks to landfills:
- Biodegradable waste such as paper and yard waste cannot completely break down.
- Methane gas is a product of anaerobic decomposition and must be managed as its generated in a landfill.
- Leachate, liquids in waste, must be managed so that it does not contaminate groundwater and nearby surface water.

Modern landfills are built with liners and methane collectors to prevent some of these problems. Unfortunately, thousands of old and abandoned landfills do not have such systems.

Incinerators

Incinerators are facilities that burn trash at high temperatures to reduce the volume of waste. Approximately 16% of MSW is burned in incinerators across the country. Most of MSW is burned in *mass-burn incinerators*, in which mixed trash is burned without sorting of hazardous and noncombustible materials. Incinerators are very expensive to build, operate, and maintain. Incinerators can emit air pollution and after burning trash, there is ash remaining that must be dealt with. Usually it is buried in a landfill.

Reduce, reuse, and recycle is the best way to deal with solid waste. Scientists estimate that about 75% of solid waste produced today could be eliminated through the "three R's" strategy.

Reduce

A key factor in reducing solid waste is to *decrease consumption* of materials. Ask yourself if you really need the product. Individuals can reduce hazardous waste by using less hazardous cleaning products. Baking soda, vinegar, and borax are inexpensive and nonhazardous products that can be used for cleaning. Industries can reduce waste by redesigning manufacturing processes to use fewer materials and to produce less pollution.

Reuse

We are in a "throw away society" and few people think of reusing a product if it is cheap and easy to obtain. Individuals can use cloth napkins and towels instead of disposable paper products. Also, you could take reusable cloth bags for shopping at the grocery store.

Recycle

Recycling is a simple word for a complex process. A product is used; it is then collected and made into a new product. Another consumer then purchases the new product made of recycled material.

Most materials are recyclable:
- Paper
- Cardboard
- Glass
- Aluminum cans
- Metal food cans
- Plastics

Composting

Another strategy to reduce, reuse, and recycle is to compost biodegradable material. Material such as food scraps, yard waste, and paper can break down naturally in the environment with the help of organisms such as worms, bacteria, and fungi. The end result is **compost**, which is a nutrient rich soil that can be used in flower beds and gardens.

organic material + oxygen + water + decomposing organisms = compost

If you have a backyard,
Dig a shallow hole to place the organic waste in. Once a week, mix the material and turn it over. In a few months, nature will take its course and compost will form.

References:

BrainyQuote. (2014). Dan Phillips. Retrieved from http://www.brainyquote.com/quotes/keywords/landfill.html

Environmental Protection Agency. (20125. Municipal solid waste.. Available at http://www3.epa.gov/epawaste/nonhaz/municipal/index.htm

Shah, Kanti L. (2000). *Basics of solid and hazardous waste management technology*. New Jersey: Prentice Hall.

I. Garbage Assessment
Procedure:

1. Work in teams as indicated by your instructor.
2. Obtain a bag of sample MSW.
3. Sort and analyze the waste; complete the table below.

Waste Category		Types of items included	Waste weight	Waste % Volume	Recyclable at Orange Grove?	
					Yes	No
Paper						
Food Waste						
Metal						
Glass						
Plastics						
Other						

4. Recyclable and Non-recyclable Materials
Based on items that the "Orange Grove Recycling Center" recycles, separate your MSW into recyclable and non-recyclable items.

Recycling Guide for Orange Grove Recycling Center
Glass Containers
Can recycle any clear, green, brown, or blue container that is glass.
Cannot recycle: window glass, light bulbs, mirrors, ceramics, dinnerware, drinking glasses.
Metals
Can recycle aluminum cans and steel food cans.
Cannot recycle aluminum foil, pie plates, or wire hangers.
Plastics
Can recycle #1 (soft drink bottles), #2 (milk jugs, shampoo/detergent bottles), and #3-#7.
Cannot recycle plastic toys, styrofoam, automotive fluid or chemical bottles.
Paper
Can recycle newspapers, glossy ads, junk mail, catalogs, books, magazines, copier paper, notebook paper, and boxes. Cannot recycle paper that is contaminated with food.

SOLID WASTE DATA SHEET

Name:_____

Answer the following questions based on your sorted garbage and data compiled in the table.

1. Which waste category was the largest component by weight?_____

2. Which waste category was the largest component by volume? _____

3. Which of the recyclable items have the "chasing recycle arrows" on the product or label indicating that it can be recycled?

_____

_____

4. Which of the items have a symbol or word indicating it was made of recycled materials? _____

_____

5. Reply to the following statements concerning your lifestyle in relation to the solid waste stream.

Statement	Yes	No	I plan to do
I buy only what I need.			
I take my own shopping bag to the grocery store.			
I get plastic bags at the store and reuse them at home.			
I use cloth napkins and towels instead of paper.			
I buy products made from recycled materials.			
I compost my yard and food wastes.			
I use refillable mugs and water bottles.			
I use a lunch box and reusable plastic containers.			
I use rechargeable batteries.			
I buy products with minimal packaging.			
I recycle aluminum cans.			
I recycle glass.			
I recycle paper.			
I recycle plastic.			

II. Landfill Assessment
Perform the described calculations to determine the size of landfill needed.

Your county's sanitary engineer has been asked to meet with the mayor and city manager to discuss disposal of solid waste. A new landfill will be opened in the Northwest part of the county and the mayor wants to know how much land will be needed to accommodate waste for 100,000 residents over the next 20 years.

Step 1:
The national average of per capita solid waste production is 4.5 pounds a day. How much waste will each person produce in one year? _____

Step 2:
Approximately 34% of solid waste is recycled and composted and the rest will be sent to the landfill. How much waste per person per year will be sent to the landfill? _____

Step 3:
Considering the landfill will serve 100,000 residents, how much total waste will be sent to the landfill in one year? _____

Step 4:
How much solid waste will the residents send to the landfill in 20 years?

_____

Step 5:
The sanitary engineer estimates the waste will be compacted at the landfill by bulldozers to a density of 1000 lbs/yd^3. What is the total volume needed for the waste calculated in step 4?

_____

Step 6:
The engineer's knowledge of the area's geology indicates that bedrock occurs at 12 feet below the soil surface. Therefore, the landfill will be 12 feet deep. Calculate the area needed for the landfill.
a. Convert volume (step 5) from cubic yards to cubic feet. 1 cubic yd = 27 cubic ft

_____

b. Calculate area (volume/depth) _____

Step 7: Convert to acres. One acre = 43,560 feet.

_____